Multifractals and Chronic Diseases of the Central Nervous System

Dipak Ghosh • Shukla Samanta
Sayantan Chakraborty

Multifractals and Chronic Diseases of the Central Nervous System

Springer

Dipak Ghosh
Department of Physics
Sir C V Raman Centre for Physics and
Music, Jadavpur University
Kolkata, West Bengal, India

Shukla Samanta
Department for Physics
Seacom Engineering College
Howrah, West Bengal, India

Sayantan Chakraborty
Electrical and Electronics Engineering
ICFAI University
Agartala, Tripura, India

ISBN 978-981-13-3551-8 ISBN 978-981-13-3552-5 (eBook)
https://doi.org/10.1007/978-981-13-3552-5

Library of Congress Control Number: 2018964106

This Springer imprint is published by the registered company Springer Nature Singapore Pte Ltd.
The registered company address is: 152 Beach Road, #21-01/04 Gateway East, Singapore 189721, Singapore

Preface

"Neurodegeneration" corresponds to any pathological condition which primarily affects neurons. Neurological disorders not only affect the brain but also the nerves that are found throughout the body and spinal cord. Neurodegenerative diseases (NDDs) are defined as disorders that affect the central nervous system causing progressive dysfunction of the nervous system. These incurable and exhausting conditions are characterized by loss of neuronal cell function and are often associated with the deterioration of structures of the affected nervous system. Diagnosis of neurological diseases is a growing concern and one of the most difficult challenges for modern medicine. According to the World Health Organization's recent report, neurological disorders, such as epilepsy, Alzheimer's disease, Parkinson's disease, Huntington's disease (HD), stroke, and headache, to name a few, affect up to one billion people worldwide. An estimated 6.8 million people die every year as a result of neurological disorders.

The book primarily focuses on the study of different neurological disorders like epilepsy, Alzheimer's, Parkinson's, HD, and motor neuron diseases (MNDs) from a new perspective by analyzing the physiological signals such as EEG, EMG, ECG, and gait rhythm associated with these diseases using nonlinear dynamics.

Physiological signals such as heart rate, blood pressure, respiration, and stride intervals fluctuate continuously over time reflecting the complex regulation of these signals by the central nervous system. Analyzing the dynamics of these human physiological signals is an important area of research to help control and to be able to predict the onset of pathological conditions. Since long, several researchers have used different techniques, mostly linear, for studying various diseases, but, of late, research on nerve-related diseases or disorders has gained much importance as the world suffers a lot of deaths due to these progressive neuron diseases which are a slow and silent killer due to the lack of proper knowledge and appropriate medication. Since we have been working on various neurological disorders for more than 10 years, we decided to summarize our works and also the works of other researchers in this field in the form of a book.

Like every other system found in nature, physiological signals are also of complex character, as they are composed of many subsystems which are strongly correlated to each other, but not in a linear fashion. Conventional linear techniques like amplitude, root mean square, or Fourier analysis cannot provide detailed information about these subsystems. The development of nonlinear methods has significantly helped in studying complex nonlinear systems in detail by providing accurate and precise information about them. Nonlinear time series analysis methods enable the determination of characteristic quantities of a particular system solely by analyzing the time course of one of its variables. Thus, from this viewpoint, nonlinear time series analysis methods are superior to mathematical modeling, since they enable the introduction of basic concepts directly from the experimental data. Since works of several researchers have established the complex nonlinear character of physiological signals like EEG, ECG, EMG, and human gait rhythm, we have been motivated to use nonlinear techniques in our work.

In a nutshell, the book provides a comprehensive study on most of the neurological disorders with special emphasis on the methods used which are not only new but also rigorous and robust. The findings provide simple parameters for the diagnosis and prognosis of different neurodegenerative disorders, and adequate software can be developed which can easily be coupled with machines. The overall premise of the book for analyzing bioelectrical signals using nonlinear techniques is easily achievable and need of the day. The book will be easily accessible and useful to a very large community working in biomedical sciences and engineering.

We hope that this compilation of the original research work on the analysis of brain through fractal analysis will definitely provide a platform and a direction for inquisitive students and researchers of biomedical engineering and neuroscientists to think objectively on the premises. We also feel that this work will stimulate more exhaustive research on different other neurological disorders which are not commonly studied.

We are really happy that the leading publisher Springer Nature has accepted to publish the book. We sincerely thank the editors and all the other staffs of Springer Nature for their continual help, support, and suggestions.

Kolkata, West Bengal, India — Dipak Ghosh
Howrah, West Bengal, India — Shukla Samanta
Agartala, Tripura, India — Sayantan Chakraborty

Acknowledgment

We begin this acknowledgment section with Albert Einstein's philosophical thought:

> *Many times a day I realize how much my own outer and inner life is built upon the labors of my fellow men, both living and dead, and how earnestly I must exert myself in order to give in return as much as I have received.*

We would like to express our deepest appreciation to all brilliant authors, researchers, and doctors who have provided us with a wealth of wisdom in the domain of diseases of the central nervous system. We believe that the analysis of biomedical signal with a novel new approach yielding precise results has become possible due to a breakthrough idea of Mandelbrot introducing a new concept of "fractals." We take this opportunity to record our acknowledgment to Benoit Mandelbrot.

We would like to express our heartiest thanks to Springer Nature and the editors for their kind support and encouragement which have helped us in the completion of the book. We would also like to express our gratitude toward the reviewers for their kind cooperation and important suggestions which have helped to enrich the book. We also want to extend our thanks and appreciation to all the other staff members of Springer Nature for their continual support and suggestions.

Contents

About the Authors

Prof. Dipak Ghosh a University Gold medalist and a Ph.D. in High Energy Physics, is currently an Emeritus Professor at Sir C.V. Raman Centre for Physics and Music, and Director (Hony.) of the Biren Roy Research Laboratory for Radioactivity and Earthquake Studies, Jadavpur University. Formerly, he was a Professor and Dean of the Faculty of Science, Jadavpur University. Prof. Ghosh has received several awards like the S.N. Bose Memorial Medal, C.V. Raman Award by the University Grants Commission, etc. in recognition of his in-depth research work. He was the Principal Investigator from India in projects with prestigious research institutes like CERN and FERMI Lab. He has over 500 publications in reputed international journals and several books to his credit. His research interests include nuclear and particle physics, nonlinear dynamics applied to biomedical problems, and neurocognition of music. He has also served as a reviewer for international journals like *Physical Review Letters*, *Physica A*, *EPL*, etc.

Dr. Shukla Samanta who holds a Ph.D. in Physics from Jadavpur University, is currently associated with Seacom Engineering College, Kolkata, as an Assistant Professor and Head of the Basic Science and Humanities Department. She holds a postgraduate degree in Physics from Presidency College, and an M.Tech. in Computer Science and Application from the University of Calcutta. She has published several papers on seismic activity, econophysics, and physiological systems in various peer-reviewed international journals. She has also served as a reviewer for international journals like Physica A, IJAV etc. Her current research interest is focused on the nonlinear study of different neurological disorders.

Mr. Sayantan Chakraborty is currently pursuing his Ph.D. in Electrical and Biomedical Signals at the Department of Instrumentation and Electronics Engineering, Jadavpur University. He completed his postgraduate degree in Electrical Power Engineering at the University of Calcutta. He has published several papers on electric bid prices, the Stock Exchange of India and biomedical signals in journals of international repute. He has also served as a reviewer for international journals such as *Physica A*, *IET Science, Measurement & Technology*, etc. Mr. Chakraborty is also an Assistant Professor and Head of the Department of Electrical and Electronics Engineering at ICFAI University, Tripura. He was previously affiliated with Dr. Sudhir Chandra Sur Degree Engineering College and Seacom Engineering College, Kolkata, as an Assistant Professor and Head of the Department of Electrical Engineering.

List of Figures

List of Tables

Chapter 1
Introduction

Abstract Disease of the central nervous system has been described in the literature as a group of neurological disorders for which the function of the brain or spinal cord is affected. This chapter outlines the general description of the diseases like epilepsy, Parkinson's, Huntington's, Alzheimer's, and motor neuron diseases. Also a discussion on the diagnostic tools and the methodologies adapted is reviewed in detail.

1.1 Central Nervous System and Its Diseases

The nervous system controls all activities of human beings. The nervous system consists of the brain and the spinal cord, as well as all the nerves throughout the body. Compared to other living organisms, humans are considered to be superior as the anatomy and physiology of the nervous system of humans are unique. The brain and spinal cord form the central nervous system (CNS), and all other nerves throughout the body are referred to as the peripheral nervous system (PNS). As the central nervous system (CNS) can determine the consciousness of us humans, it has been attributed to be the most complex organ in the human body. All aspects of our behavior from breathing to supporting our thoughts and feelings (Kandel and Squire 2000) are controlled by the nervous system. The human brain is the most sophisticated organ in the human body. The brain regulates vital body functions such as emotion, memory, cognition, motor activities, heart rate, respiration, and digestion. The human brain is a complex network of millions of neurons packed in a matrix of glial cells (Benson et al. 2017).

Diseases of the brain may be caused either due to inherent dysfunction of the brain or due to complex interactions of the brain with the environment (Hyman et al. 2006). Brain diseases range widely from common neurological to psychiatric disorders. Throughout the life span, brain diseases affect a very significant portion of the population and are widely spread both across the developed and developing nations. Compared to other diseases, brain diseases account for the highest burden in terms of health, economy, and social capital globally (Nathan et al. 2001). More than 1.5 billion people are affected due to brain disorders worldwide, and with the passage of time, it is feared that this population will increase. Thus there is an urgent need of not

D. Ghosh et al., *Multifractals and Chronic Diseases of the Central Nervous System*,
https://doi.org/10.1007/978-981-13-3552-5_1

only producing more drugs to treat CNS disorders but rigorous research so that early prognosis and diagnosis can be made and is the need of the day to help control this epidemic.

Across the life span of human, the nature of the brain disorders changes. In young there is a high prevalence of psychiatric disorders, like depression, anxiety, schizophrenia, and substance abuse, whereas the elderly suffer markedly from neurodegenerative disorders such as dementia or stroke (Wittchen et al. 2011). More widely appreciated are the neurodegenerative disorders, namely, Parkinson's (PD) and Alzheimer's disease (AD), which are on the rise due to an older population (von Campenhausen et al. 2005). Huntington's disease (HD) and amyotrophic lateral sclerosis (ALS) are other neurodegenerative diseases. The neurodegenerative diseases are characterized by inevitable gradual decline in cognitive ability and also the potential to self-sustain (Prince et al. 2013). The neurodegeneration produces a clinical syndrome called dementia, whose symptoms include inability to recollect, sudden and unexpected changes in one's mood, and problems to communicate and reason (Devous 2002).

Next to stroke, epilepsy is the second most common neurological disorder affecting approximately 50 million people worldwide. "Epilepsy" is derived from the Greek term *epilambanein* which means to seize, and it denotes the predisposition to have recurrent, unprovoked seizures (Quintero-Rincon et al. 2016). In epilepsy, the nerve cells of the brain transmit exorbitant electrical impulses that cause seizures. An epileptic seizure is defined as "a transient symptom of excessive or synchronous neuronal activity in the brain" (Fisher et al. 2005). Epilepsy is defined by two or more such unprovoked seizures. Seizures may be either focal or generalized. In focal seizures, only a specific segment of the brain is affected, while in generalized seizures the whole brain is affected (Acharya et al. 2012a). Epileptic seizures may lead to impairment or unconsciousness and psychic, autonomic, sensory, or motor problems (Lehnertz 2008).

Electroencephalography (EEG) is an important clinical tool for monitoring and diagnosing neurological changes in epilepsy. Compared with other methods such as magnetoencephalography (MEG) and functional magnetic resonance imaging (fMRI), EEG is an affordable and safe technique for inspecting brain activity. Drugs and surgical treatment options are not sufficient to treat epilepsy. Among new therapies developed, implantable devices that deliver direct electrical stimulation to affected areas of the brain have shown promising results. The effectiveness of these treatments depends mainly on robust algorithms for seizure detection. As seizure onset cannot be predicted, a continuous recording of the EEG is essential to ascertain epilepsy. But since visual assessment of long EEG recordings is tedious and time-consuming (Song 2011), automated detection methods of epilepsy have gained importance. With a view to study the changes that occur in the brain in seizure and seizure-free status, we analyzed EEG signals using a latest state-of-the-art methodology. The observations made are very interesting and are described in Chap. 2.

Alzheimer's disease another disease of the central nervous system is the major reason of dementia. This disease is described by an intensifying reduction in brain function, which generally commences with deterioration in memory (Devous 2002). Alzheimer's disease progression can be grouped into four stages. The "preclinical" stage is mild cognitive impairment (MCI) where patients show mild memory impairment, but their other cognitive abilities and functional activities are retained (Petersen et al. 2009; Weiner et al. 2012). In the next stage, reduction of independence is caused due to increasing cognitive insufficiency. Mild AD and moderate AD are the second and third stage of Alzheimer's disease. In severe AD, the last stage, not only are all cognitive functions acutely damaged, but also chewing and swallowing become extremely difficult (Bianchetti and Trabucch 2001). Early diagnosis of AD in MCI and mild AD stages is important so that with proper medical intervention, severity of the disease can be checked.

Neuroimaging techniques, physiological markers, and genetic analyses are the methods used for the detection of Alzheimer's disease. Popular neuroimaging techniques used successfully in AD detection at an early stage include single-photon emission computerized tomography (SPECT), positron emission tomography (PET), and magnetic resonance imaging (MRI) (Elgandelwar and Bairagi 2016). But the problems with these methods are that they are not only expensive, time-consuming, and inconvenient but also possess risks due to radiation. In order to understand the macroscopic spatial temporal dynamics of the electromagnetic fields of the brain, EEG and MEG have evolved as significant neurophysiologic methods (Stam 2005). In particular, EEG provides different time series leads revealing the working of the brain corresponding to different cognitive states involved in processing the information (Ronghua et al. 2001). The brain being a non-linear system, its behavior cannot be merely decomposed into single neuron behavior (Aike and Huiming 1994). The studies conducted to understand the brain dynamics in Alzheimer's disease patient have been outlined in Chap. 2 where mainly the use of EEG signals has been highlighted.

Some studies have recorded that by analyzing the activities of the autonomic nervous system (ANS), epileptic seizures can be detected. Heart rate variability (HRV) analyses can provide information of the underlying activities of the ANS. Time interval measurement between subsequent QRS complexes reflects heart rate variability which indicates the governance of the heart rate by the ANS by means of its sympathetic and parasympathetic control mechanisms (Ponnusamy et al. 2012). Thus indirect evidence of nervous system activity can be demonstrated by HRV analyses (Pavei et al. 2017). Electrocardiographic (ECG) signals have been used by some authors (Malarvili and Mesbah 2009; Varon et al. 2014; Fujiwara et al. 2016) for detecting and predicting seizures. Few studies have also focused on the study of HRV parameters in subjects before seizure and in between seizures (Ponnusamy et al. 2011, 2012; Behbahani et al. 2013; Pavei et al. 2014). But much work has not been done to assess autonomic nervous system (ANS) activity post seizure. A study on post-seizure heart rate oscillation is outlined in Chap. 3.

Amyotrophic lateral sclerosis (ALS) is a motor neuron disease. Problem to speak, swallow, breathe, and contract and relax muscles, stiffening of muscles, progressive weakness, etc. are some typical features associated with this disease. Damage of motor neurons in any portion of the body is affected by amyotrophic lateral sclerosis disease. Since the sensory nerves and involuntary nervous system are not affected in most of the patients, the sensations of touch, hearing, smell, sight, and taste (Singh et al. 2013) are intact. Several investigators have studied the stride interval to detect ALS (Yunfeng and Sin 2010; Sugavaneswaran et al. 2012). Hausdorff et al. (1985) compared the gait rhythm of patients suffering from ALS with healthy subjects and subjects suffering from Parkinson's disease (PD) and Huntington's disease (HD). In Chap. 4, we have described motor neuron disease (MND) which is also a neurological disorder of the motor system in adults. Since motor neurons are the nerve cells that are responsible for movement of the muscles, thus they are considered to be the most significant part of the neuromuscular system. As MND disease damages motor neurons in the brain and spinal cord, the muscles gradually become weaker, and after a certain period, it stops working. Electromyography (EMG) is the most important diagnostic tool of MND. In Chap. 4 using a multifractal approach, we have briefed the analysis of EMG signals of people suffering from myopathy and neuropathy.

Parkinson's disease (PD) is an incurable and advancing hypokinetic disorder of the central nervous system (Jankovic 2008). In 1817 James Parkinson named it "the shaking palsy." Though people over 50 years are seen to be affected by the disease where symptoms evolve gradually over time, some rare form of the disease where progression rate is faster affects the younger ones. Parkinson's disease is not only chronic and progressive, since it prevails over a longer time period, but the symptoms become worse as time progresses (Singh et al. 2013). It is caused by basal ganglia dysfunction. Basal ganglia, which emanate from the cerebral cortex and the brain stem, produce motor impulse and send sensory information through the projecting loops in the central nervous system (Sian et al. 1999). Functioning of the motor system is affected by basal ganglia dysfunction which leads to impaired balance and altered gait rhythm (Marsden 1982). Thus study of gait rhythm is of immense importance in understanding Parkinson's disease. A detailed study of gait dynamics using non-linear approach is presented in Chap. 5.

Huntington's disease (HD) is an autosomal-dominant neurodegenerative disease of the central nervous system. The disease derived its name from George Huntington. It was first described by him in 1872 hence the name (Hausdorff et al. 1997). Chorea and cognitive and personality changes are the initial clinical features. Pathological changes are mostly seen in the basal ganglia, with a loss of neural projection in the striatum (caudate nucleus and putamen) (Penney and Young 1993). People in the age group of 30 and 40 who have generally not been affected by concomitant disease and age-related physiological changes are commonly affected by Huntington's disease. Essentially Huntington's disease is a manifestation of some nonfunctioning of central nervous systems. It provides a suitable contrast to aging, useful in studying dynamics behind stride-interval correlation with nonfunctioning of the brain. The capability of the locomotor system to generate correlations in the stride interval was shown to be decreased in elderly subjects and in subjects with HD by Hausdorff et al. (1997). To test this theory, we have studied the fluctuations of gait dynamics of HD and PD and compared them with healthy controls. The results of the study are contained in Chap. 5.

Alterations in heart rate and blood pressure due to change in body posture are an important area of research to understand orthostatic stress. However no detailed study has been done and reported in existing literature. Multifractal cross-correlation study between the two physiological signals, namely, ECG and arterial blood pressure (ABP) described in Chap. 6, highlights the responses in the cardiovascular system due to changes in posture.

1.2 Bioelectrical Signals

Due to the complex nature of the nervous system, neurological disorders are the most challenging to diagnose, manage, and monitor. Diagnosis and treatment require high precision, dedication, and experience. Different disorders of the nervous system include epilepsy, dementias, Alzheimer's disease, cerebrovascular diseases including stroke, multiple sclerosis, Parkinson's disease, migraine, neuroinfections, brain tumors, and traumatic disorders of the nervous system to name a few. Structural, biochemical, or electrical abnormalities in the brain, spinal cord, or other nerves give rise to a wide variety of symptoms, thus developing various disorders. Population worldwide regardless of their sex or age is afflicted by these disorders, thus facing devastation and deprivation. So a timely and speedy diagnosis of these disorders is necessary to prevent and significantly improve the quality of life by acquiring suitable technologies. Wide variety of modern diagnosis methodologies are adapted to help ascertain, control, and treat neurological disorders, such as brain wave tests (electroencephalography or EEG), computerized tomography (CT scan), magnetic resonance imaging (MRI scan), electromyography (EMG), arteriogram (also called an angiogram), and positron emission tomography (PET scan or PET imagery). With these important tools, the physicians can affirm or adjudicate the existence of a neurological disorder or other medical conditions. These diagnosis techniques produce huge quantity of data which is analyzed manually hence time-consuming to detect and is also prone to error and fatigue. It is a difficult task to assemble, supervise, and understand such huge data inspecting visually so computerized diagnosis techniques have been developed which can reliably detect the neurological abnormalities from these large medical data easily. Thus computer-aided diagnosis techniques would improve consistency of diagnosis and escalate the outcome of treatment, saving lives and reducing cost and time as well (Siuly and Zhang 2016). This book addresses the problem of handling big medical data using non-linear approach to quantify different disorders associated with the central nervous system. In this section we have focused on mainly the bioelectrical signals such as electroencephalography, electrocardiography, and electromyography (EMG). Later analysis of these signals in order to understand the disorders related to the brain, the cardiovascular system, and the motor dysfunction is detailed.

1.2.1 Electroencephalography

Electroencephalography is defined as the technique of recording electrical activity due to ionic current flows generated by neurons in the brain (Libenson 2009). EEG provides very useful information on the exploration of brain activity, diagnosis of cerebral disease, and identification of brain disorders (Sheng and Chen 2011). In 1875 Richard Canton was the first to identify electrical currents in our brain. Hans Berger a German neuropsychiatrist first recorded these currents and called it electroencephalogram (Berger 1929). Since the first EEG potentials were discovered, it has found its potential use in many clinical settings such as the diagnosis of schizophrenia (Schellenberg and Schwarz 1993) and in brain dynamics research in different physiological states of the patients.

Complexity, non-linearity, nonstationarity, and random nature of the EEG signals are a result of complex interconnections between billions of neurons. The signal consists of a number of sinusoidal components of distinct frequencies communicating non-linearly to produce one or more sinusoidal components at sum or difference frequencies (Zhuo et al. 2008). The non-linear character of EEG signals may unveil the hidden complexities present in the time series. In a primeval study by Babloyantz et al. (1985), non-linear parameters like correlation dimension (CD) and largest Lyapunov exponents (LLE) were used to study the sleep wave signal. Because of such studies, the probable application areas of EEG non-linear analysis methodologies have diversified. Among other applications of non-linear methods, detection of epilepsy is the most common area of application (Mormann et al. 2003, 2005; Lehnertz 2008). Detection of epilepsy from EEG signals using non-linear methods has been the focus of numerous investigations thus quantifying non-linear mechanisms thereby providing a better reflection of the EEG characteristics (Acharya et al. 2012a). EEG signals are also helpful in detecting other neurological disorders like Alzheimer's, Parkinson's, Huntington's, autistic spectrum disorder (ASD), brain tumor, etc.

1.2.2 Electrocardiography

An electrocardiogram (ECG) is a bioelectrical signal that records the heart's electrical activity against time. For assessing the heart functions and its disorders, ECG is the most important diagnostic tool. While recording ECG the electrical impulses while the heart is beating are recorded. The record manifests if there is any problem with the rhythm of the heart (Vanage et al. 2012).

A typical ECG signal is shown in Fig. 1.1. The time between two successive beats is known as RR interval. The natural pattern of the human heart is known to differ with sympathetic and parasympathetic signals. Normally, sympathetic activity is known to decrease the RR intervals and its variability, while parasympathetic activity is known to increase the RR intervals and its variability (Berntson et al.

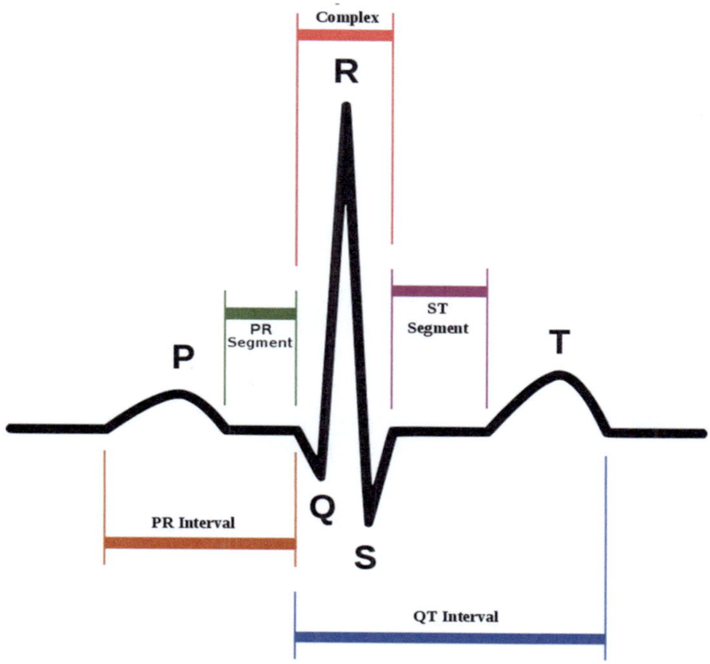

Fig. 1.1 A schematic diagram of ECG. (Source: https://en.wikipedia.org/wiki/QRS_complex)

1997). Heart rate variability (HRV) is a measure of the beat-to-beat variability (RR intervals) which indicates the health of the cardiovascular system. It is generally used to determine the coordination of cardiac autonomic function (Flynn et al. 2005; Vinik et al. 2013). A variety of linear, non-linear, periodical, and nonperiodical oscillation patterns are present in heart rate fluctuations (Aubert et al. 2002).

Among the four chambers of the heart, the upper two chambers are called the atria, and the lower two chambers are called the ventricles (Al-Qazzaz et al. 2014a). To identify ECG features, the letters PQRST are normally used. Activation of the atria is expressed by the P wave which is the first wave of the cardiac cycle. Conduction of the cardiac impulse proceeds from the atria through a series of specialized cardiac cells (the AV node and the His-Purkinje system). Following the P wave, there is a relatively short isoelectric system. A quick and huge deflection is observed on the surface of the body when large muscle mass of the ventricles is excited which causes contraction of the ventricles yielding the required force necessary to circulate blood to various organs of the body. This large wave has many components where the first downward deflection is known as the Q wave, the upward deflection the R wave, and the terminal downward deflection the S wave. Based on the location of the leads on the body and any existent abnormalities, these three waves are present. The large ventricular waveform is called the QRS complex which is succeeded by another relatively short isoelectric segment. After this short segment, the ventricles return to their electrical resting state, and a wave of repolarization is seen as a low-frequency signal known as the T wave (Bronzino 2000).

HRV measurements exhibit non-linear and nonstationary character, and the dynamics of their fluctuation with varying time periods contains significant information (Ivanov et al. 1999, 2001, 2004). With the use of conventional (linear) mathematical methods for analyzing fluctuations in heart rate variability, important information gets lost. So development and implementation of new non-linear mathematical methods based on the fractal theory will govern the cause of HRV fluctuations. HRV signals possessing properties of fractal geometry like self-similarity, scalability, fractal dimension, and long-range dependency are accounted for the basis of development of non-linear methods (Stanley et al. 1999; Rhaman et al. 2013). Application of non-linear methods is of extreme importance because the results will present intricate report of physiological status of the patients. Another potentiality of these non-linear methods is acquaintance of new knowledge concerning the diagnostics, prognosis, and prevention of pathology of cardiovascular disease (Gospodinova 2014).

1.2.3 Electromyography

The electromyography (EMG) signal is a bioelectrical signal which records electrical currents generated in muscles during its contraction portraying neuromuscular activities. Contraction or relaxation of muscles is governed by the nervous system. Hence, EMG is a complex signal, which is controlled by the nervous system and is dependent on the anatomical and physiological features of muscles (Reaz et al. 2006). EMG is of two types, namely, intramuscular electromyography (IEMG) and surface electromyography (SEMG) (Farina and Negro 2012). Intramuscular electromyography (Monsifrot et al. 2004) is an invasive technique which is done by inserting needle electrodes through the skin into the muscle tissue and the electrical signal is read by a trained professional, whereas electrical activity recording of muscles by placing electrodes on skin surface is known as SEMG (Merletti and Farina 2008; Soo et al. 2009; Hug 2011). The surface EMG signals have found considerable use in determination of motor function and movement disorders in humans, thus presenting significant information on neuromuscular strategies (Farina et al. 2004). Analysis of muscle fatigue has been done using SEMG signals (Venugopal et al. 2014). These signals are aggregate of electrical activity of the muscles and crop up as fluctuations that are vastly nonstationary and non-linear and exhibiting fractal characteristics (Talebinejad et al. 2009; Zhang et al. 2010; Meigal et al. 2013). EMG signals can detect abnormal electrical muscle activity in amyotrophic lateral sclerosis (ALS or Lou Gehrig's disease), carpal tunnel syndrome, muscular dystrophy, sciatic nerve dysfunction, inflammation of muscles, etc.

Apart from those mentioned above (EEG, ECG, and EMG), there are other recently developed advanced diagnosis technologies that can detect, manage, and treat neurological disorders such as computerized tomography (CT scan), magnetic resonance imaging, arteriogram (also called an angiogram), and positron emission tomography (PET scan or PET imagery). EEG, ECG, and EMG tools are more preferred than CT, MRI, angiogram, and PET as they are easy to detect and affordable. Where the world still suffers from hunger and poverty at large, affordable and cost-effective medical tools still serve as the primary diagnostic tools to the population worldwide. The different bioelectrical signals discussed so far have potential use in determining various physiological and pathological disorders of the nervous system, cardiovascular system, and neuromuscular system as well. Our book primarily focuses on the extraction of valuable information from these signals using non-linear signal analysis techniques.

1.3 Linear Signal Processing Techniques

Signal processing is the process of modeling, detection, and identification of patterns and structures in a signal. A signal describes the information through variation of quantity which reflects the properties, the characteristics, the state, the course of action, and the information about a source, and that information may be processed directly by humans or machines for the purpose of decision, forecasting, control, investigation, research, and further exploration of an object (Vaseghi 1996).

Physiological systems belong to a class of very complex systems. Compared to complicated systems which can be split into pieces, examined fragment wise, and reorganized back together, the complex ones are not merely sum of their parts (Jovic and Bogunovic 2010). They are non-linear to a degree and can never be inferred completely (Goldberger 1996). The prime method to explain complex systems is to represent them with a complicated model. Determination of physiological functions of a specific biological system is based on diverse clinical tests and measurements. The brain, heart, lungs, and nervous and muscular systems are the vital physiological systems in human body. Electrical currents transmitted from biological systems are due to the electrolytic activity in tissue cells which are determined by placing electrodes either on the skin or injecting them deep into the tissue (Jovic and Bogunovic 2010). Acquisition of vital information from biological and physiological systems is the key to biomedical signal processing. Since these signals help us to extract information about the current state of health of humans, their monitoring and interpretation have significant diagnostic value for clinicians as well as for researchers to understand human health and diseases.

Signal processing techniques can be classified into linear and non-linear methods. A signal is said to be linear when output is always proportional to input, whereas it is non-linear if output and input are not proportional.

Linear signal processing techniques include mainly the root mean square (RMS) method, short-time Fourier transform (STFT), fast Fourier transform (FFT), wavelet transform (WT), and discrete wavelet transform (DWT).

1.3.1 Root Mean Square (RMS) Method

RMS is the most familiar and also the easiest method for classifying signals. Root mean square values designate the variations of normal signal. For simple events, this method gives satisfactory results, but for complex system involving nonstationarity and non-linearity, this method is not suitable. Its disadvantage is its reliance on the lower signal size and its deficiency to discriminate between fundamental frequency and harmonics.

1.3.2 Fast Fourier Transform (FFT) Method

Fourier transform is a method that converts a time domain signal into its corresponding frequency components. This method is productive for periodical signals to obtain its magnitudes and phases. The discrete Fourier transform (DFT) is an advanced technique of Fourier transform analysis, and FFT is regarded as a faster version of DFT, where sampling is done in a windowed manner. For analyzing the nonstationary and non-linear signals, FFT is considered as a suitable method. Since FFT can decrease the overall calculation time, an effective mitigation algorithm can be chosen (Augustine et al. 2016).

FFT is not efficient in the analysis of nonstationary and non-linear signals. The introduction of FFT can reduce the overall computation time and thereby able to choose an efficient mitigation algorithm (Augustine et al. 2016).

Several automated EEG detecting methods have been developed, based on Fourier spectral analysis assuming EEG signals are stationary permitting signal transformation from time to the frequency domain. These methods permit researchers to gather information solely in the frequency domain (Polat and Gune 2007). Polat and Gune (2007) employed Fourier spectral analysis for EEG signal extraction.

1.3.3 Short-Time Fourier Transform (STFT) Method

Sliding window record of FFT is the short-time Fourier transform which shows good results in the frequency domain compared to other classification methods. For time-varying and non-linear signals, it presents inappropriate outcomes. To reduce complexity in calculation, it can divide long nonstationary signals into short segments (Augustine et al. 2016). The major difficulty with Fourier analysis in the obtained signals is its total spectrum, and it cannot be used for local analysis. Short-time Fourier transform (STFT) was used to calculate the spectrum density of EEG signals (Tzallas et al. 2009; Li et al. 2016).

1.3.4 *Wavelet Transform (WT) Method*

Wavelet transform method is used to analyze signals in varied sub-bands in a careful manner. Since wavelet transform analyzes windows of varying sizes, it is a more pliable way of time–frequency representation of a signal compared to Fourier transform. A significant feature of wavelet transform is that, at high frequencies, it presents specific time information and at low frequencies it presents precise frequency information. This feature of WT is of paramount importance as biomedical signals normally include low-frequency information with long time duration and high-frequency information with short time duration. Wavelet transform can precisely measure the transient characteristics and is confined in both time and frequency domains. But the main problem with wavelet transform-based methods is the choice of mother wavelets (Song 2011).

1.3.5 *Discrete Wavelet Transform (DWT) Method*

The discrete wavelet transform (DWT) (Mallat 2002; Addison 2002) is an appropriate technique to represent and employ signals having distinct transients. It divides the signal into its low-resolution parts and a series of details at varied resolutions. Denoising is the most probable application of the DWT. It has drawn ample attention for removing noise in biomedical signals (Quiroga and Garcia 2003; Poornachandra 2008; Phinyomark et al. 2009; Li et al. 2009; Gao et al. 2010).

Discrete wavelet transform (DWT) was used by some researchers to extract features from EEG signals (Ocak 2009; Li et al. 2011). Faust et al. (2015) used the DWT-based EEG denoising method and the feature extraction for the seizure detection and epilepsy diagnosis. The study presented the wavelet technique to be an efficient method for automatic epilepsy diagnosis using EEG signals. Hassan et al. (2016) proposed the tunable factor wavelet transform for automated epilepsy diagnosis. All these traditional approaches are not sufficient alone to provide an efficient way to characterize the complexity of EEG signals without taking into account the non-linearity of the signals.

1.4 Limitations of Linear Analysis Techniques

Linear analysis of physiological signals includes frequency analysis (e.g., Fourier and wavelet transforms) and parametric modeling (e.g., autoregressive models). Though linear methods have been successfully applied in several problems (Anderson et al. 1998; Garrett et al. 2003; Faust et al. 2010), they provide a limited amount of information as they ignore the underlying non-linearity in the signal. To describe the state of a system, linear analyses mainly emphasize on central tendencies, while non-linear analyses tender some understandings in the organization of the variability of state of a system by evaluating the persistence of certain patterns or "shift" in the regularity of the time series (Darbin et al. 2013). Conventional linear techniques cannot provide detailed information about the subsystems. The development of non-linear methods has significantly helped in understanding complex non-linear systems in detail by providing accurate and precise information.

Simple non-linear systems were found to present highly complex (Grassberger and Procaccia 1983; Parker and Chua 1989; Buczkowski et al. 1998; Kim et al. 1999; Eke et al. 2002; Sarkar and Leong 2003) and chaotic behavior as they are intensely sensitive to initial conditions, since any perturbation, no matter how minute, will forever alter the future of the systems (Ghosh et al. 2017). Eke et al. (2002) proposed that the presence of self-similarity is a typical feature of complex signals as the smaller-scale structure bears resemblance to the larger-scale structure in complex bioelectrical signals such as EEG, ECG, EMG, etc. Fractals exhibit this self-similar property (Mandelbrot 1977). Thus simple linear techniques cannot provide minute detail information about complex dynamical systems. Thus, the necessity of developing non-linear analysis techniques came into force.

1.5 Non-linear Techniques

Describing natural process using non-linear dynamics is a more realistic approach as in non-linear systems, input is not directly proportional to output, and they can be described by a wide variety of behavior that corresponds to the complexity seen in nature. Non-linear systems may be viewed as immensely stable possessing limit-cycle behavior or unstable subject to the precise nature of the non-linearity.

Extreme and persistent sensitivity to perturbation may lead to instability and under certain conditions may give rise to complex and apparently unstable behavior, an incident commonly called "chaos." As a result of the underlying dynamics, physiological systems may show random fluctuations. Important information about the dynamical state of the system may be present in these fluctuations which are not possible from traditional observations (Bishop et al. 2012). Recent developments of mathematical tools and computerized methodologies have made it possible to explore these inherently difficult systems. The application of sophisticated non-linear analysis techniques has revealed the undisclosed facts of a variety of time series data of natural processes (Ivanov et al. 1999; Goldberger et al. 2002; Krishna et al. 2003; Enescu et al. 2004; Oswiecimka et al. 2005; Wink et al. 2008; Millan et al. 2010).

Several non-linear analysis techniques have developed with the advent of time. Phase space reconstruction, recurrence quantification analysis, entropy analysis, and fractal and multifractal analysis are a few. In this book we have mainly focused on the use of fractal- and multifractal-based methods to explore disorders of the central nervous system.

Phase Space Reconstruction Method: Phase space reconstruction is a standard method for studying chaotic systems. It shows the trajectory of the system in time. A set to which a dynamic system evolves after a long enough time is called attractor. A space in which each point outlines two or more than two states of a variable of a system is known as phase space or phase diagram. Phase space dimension or reconstruction dimension is defined as the number of states present in phase space (Bogunovic and Jovic 2010).

Recurrence Quantification Analysis (RQA): A technique independent of data size, stationarity, and statistical distribution and which can detect and analyze state changes in drifting dynamic systems is referred to as the recurrence quantification analysis (Webber and Zbilut 1984). It is a technique which demands precise competence. Its proper implementation in physiology can measure its potentiality. It has the capability to look at the inner structure of the examined signal (Conte et al. 2015).

Entropy Analysis: To study the regularity of a time series, Pincus (1991, 1995) proposed approximate entropy (ApEn). A lower value of ApEn corresponds to a regular and predictable time series, whereas a higher value of ApEn represents a random time series. For evaluating average logarithmic probability, it takes a template-wise approach. The method has found profound use in studying the cardiovascular system. Due to its dependence on length of the record and consistency, it has major drawbacks (Richman and Moorman 2000; Lake et al. 2002). Sample entropy (SampEn, r, N) is the negative natural logarithm of the conditional probability that a data set of length N, having repeated itself within a tolerance r for m points, will also repeat itself for m + 1 points, not permitting self-matches (Richman and Moorman 2000). Compared to ApEn, SampEn evaluates negative logarithm of probability of the total time series (Humeau et al. 2008). The main advantage of SampEn over ApEn is that it does not depend on length of the record and also shows relative consistency (Richman and Moorman 2000; Lake et al. 2002; Al–Angari and Sahakian 2007).

To study the non-linear systems, fractal-based methods have gained much importance in comparison to other methods and have been successfully applied to a variety of systems.

1.5.1 Fractals and Multifractals

A fractal is a geometric arrangement made from a vast set of increasingly small subunits. Each small unit is an imitation of the whole, which is self-similar across length scales, i.e., over different levels of magnification (Bassingthwaighte et al. 1994). In snowflakes, clouds, mountain ranges, vegetables, coastlines, etc., self-similarity is generally noticed (Mandelbrot 1967). So this phenomenon of self-similarity is found all over nature. Self-affine fractals are those when scale of variation in one direction is different from that in another direction (Mandelbrot 1985). The concept of fractals was first introduced by the Polish mathematician

Benoit Mandelbrot. He derived the term from the Latin adjective *fractus* which means broken.

Fractal analysis provides a mathematical formalism to characterize intricate spatial and dynamical structures (Feder 1988). As fractals have fractional dimension, it can help in explaining extremely irregular objects which Euclidean geometry cannot. Fractal scaling behaviors can identify variation of time series patterns with alteration of temporal scales (Ge and Leung 2013). Long-range correlation is another striking feature of fractals which means fluctuation in the time series is related to the previous fluctuation. The change of correlations is based on power law (Namazi and Kulish 2015). Fractal methodology has found wide use in biology and medicine for studying DNA (Namazi et al. 2015), eye movement (Namazi et al. 2016a), EEG signal (Namazi et al. 2016b), bone structure (Huh et al. 2011), etc.

Fractals are of three types: one that occurs in nature, geometrical fractals those which are artificially generated, and complex fractals. In nature fractals exist everywhere. The vegetables that we eat are also examples of fractals. Below are some examples of fractals that are found in nature:

- A fern consists of branches which look similar to little ferns (Fig. 1.2); those branches in turn are made of smaller but structurally alike elements. The self-similar character of fractals also does not vary with scale, since they even appear mathematically similar at all scales of observation. At all scales of examination, rocks seem to be the same. An observer looking at a picture cannot define its scale till a figure of known size is in the picture (Brown and Witschey 2003).
- The next example is a very common vegetable the Romanesco broccoli which shows self-similar form approximating a natural fractal (Fig. 1.3).
- Another image of fractal found in nature is the quartz stone. It has triangular points which look similar if one zooms in on any part of the stone (Fig. 1.4).
- Neurons the basic building blocks of the central nervous system are an example of fractal found in the brain (Fig. 1.5).

Fig. 1.2 A fractal fern: it repeats its pattern at various scales. (Source: https://commons.wikimedia.org/wiki/File:Sa-fern.jpg)

Fig. 1.3 Romanesco broccoli: showing self-similar form approximating a natural fractal. (Source: https://en.wikipedia.org/wiki/File:Fractal_Broccoli.jpg)

Fig. 1.4 A quartz stone. (https://en.wikipedia.org/wiki/File:Unknown_Quartz_crystal_66.JPG)

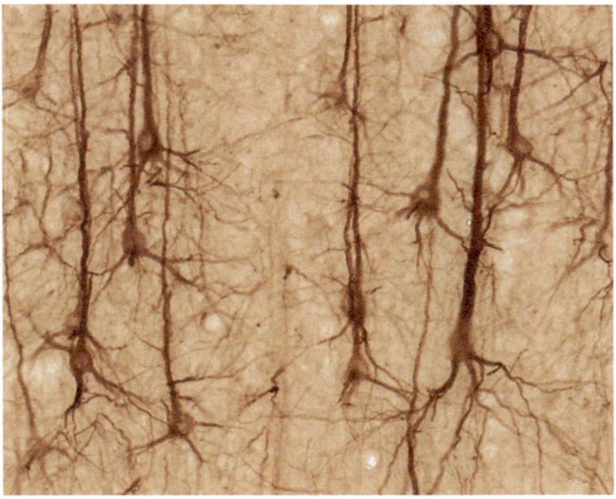

Fig. 1.5 Neurons in cerebral cortex. (Source: https://commons.wikimedia.org/wiki/File:Smi32neuron.jpg#file)

Fig. 1.6 The Cantor set. (Source: https://en.wikipedia.org/wiki/Cantor_set#/media/File:Cantor_set_in_seven_iterations.svg)

Fig. 1.7 The Sierpinski triangle. (Source: https://commons.wikimedia.org/wiki/File:Sierpinski_triangle.svg)

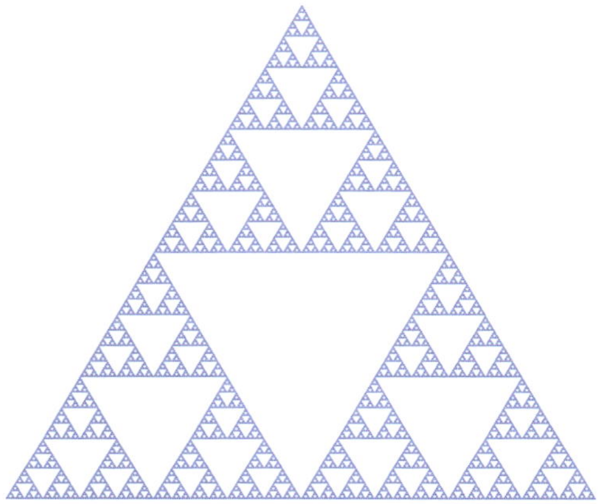

Geometric Fractals: These fractals are artificially generated structures obtained by a simple production rule. Some of the geometrically constructed fractals are Cantor set, Sierpinski triangle, Von Koch flake, etc.

(a) Cantor Set: To obtain a Cantor set, a line is first divided by 3; then its middle third part is removed. The procedure is repeated for each of the remaining parts. Continuing this way in each step, the same structure is obtained but to a smaller size (Fig. 1.6).
(b) Sierpinski Triangle: Polish mathematician Waclaw Sierpinski in 1916 described the Sierpinski triangle. In the process of connecting the middle points of each side of an equilateral triangle to produce four separate triangles and then removing the triangle in the center, we can generate the Sierpinski triangle repeating the process an infinite number of times (Fig. 1.7).
(c) The Von Koch Flake: This geometrical fractal can be constructed by dividing each straight line segment of an equilateral triangle into three equal parts and replacing the middle part with two segments of the same length. After infinite number of iterations, the Von Koch flake is produced. Self-similarity of fractals can be clearly noticed in the Von Koch flake (Fig. 1.8).

Complex Fractals

Among complex fractals, the Mandelbrot set is the most famous. Mandelbrot used the recursive formula $Z_n = Z_{n-1}^2 + C$, where C is a real number and Z is a complex number to create the Mandelbrot set. Figure 1.9 depicts a Mandelbrot set.

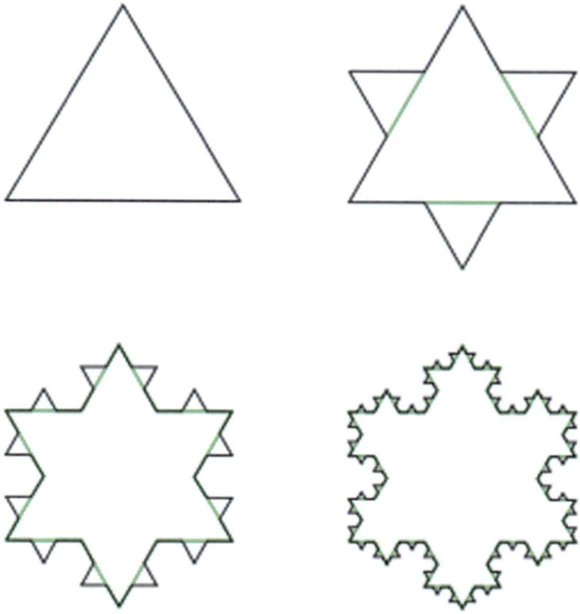

Fig. 1.8 The Von Koch flake. (Source: https://commons.wikimedia.org/wiki/File:KochFlake.svg)

Fig. 1.9 The Mandelbrot set. (Source: https://commons.wikimedia.org/wiki/File:Mandelbrot_set_with_coloured_environment.png)

Julia set: Julia sets, named after Gaston Julia (1893–1978), arise from analyzing the dynamics of complex functions. Julia sets are strictly connected with the Mandelbrot set. Julia sets can be generated using the same iterative function as that of the Mandelbrot set, the only difference being the way the formula is used. To generate a Mandelbrot set, we repeat the formula for each point C of the complex plane, always starting with $Z_{n-1} = 0$. For generating a Julia set, the value of C is kept constant during the generation process, while value of Z_{n-1} keeps changing. The

Fig. 1.10 The Julia set. (Source: https://upload.wikimedia.org/wikipedia/commons/b/b1/Julia_set_%28ice%29.png)

structure of the Julia set is decided by the value of C, or each point of the complex plane is associated with a specific Julia set (Patrzalek 2006). Figure 1.10 depicts a typical Julia set.

All the above fractals resemble self-similarity. Fractal dimension which is a value in fraction is determined by the amount of self-similarity in a fractal which is different from the conventional Euclidean dimension (Ge and Leung 2013). Felix Hausdorff (1919) was the first to define a non-integer dimension in describing monster functions. The unevenness of a time series can be subdivided repeatedly into self-similar components. Fractal size can be determined by the fractional dimension (FD). FD is the main tool to describe fractal geometry and the heterogeneity of irregular shapes (Lopes and Betrouni 2009). FD works like a magnifier, zooming and comparing different portions of the signal with the entire signal. Higuchi algorithm (Higuchi 1988) is one of the most efficient methods for calculating the fractal dimension (Simjanoska et al. 2018). There are other methods like box-counting method (Block et al. 1990), Kartz method (Kartz 1988), etc.

Fractals can be classified into two groups: monofractals and multifractals. If in different regions of the system the scaling properties are equal, they are termed monofractals, whereas for multifractals, the scaling properties are different in different regions of the system, thus making multifractals more complicated. Multifractals have differently weighted fractals which have fractional dimensions (Ghosh et al. 2004). Parameterization of fractal properties determines the behavior and mechanisms of the underlying control systems (Mandelbrot 1995; Goldberger et al. 2002).

Complex structures of living systems possess fractal-like geometry. Complex anatomical structures in the human body show self-similarity at different scales (Goldberger et al. 1990, 2002; Weibel 1991; Bassingthwaighte et al. 1994; Ayers 1997), for example, the blood vessel branching, networks of the tracheobronchial tree and neural networks in the brain, the folds of the intestine, choroidal plexus, etc. For describing the tracheobronchial tree, treelike fractals are very useful. Owing to the redundancy and irregularity (Goldberger et al. 2002), fractal structures are believed to be very stable (Zueva 2015). This stability of fractal geometry has motivated us to use it in our study.

Goldberger et al. (1990, 2002) and Ivanov et al. (1999) have reported the presence of long-range correlations in variations of healthy heartbeat. Fluctuations of stride interval of healthy human have been shown to exhibit scale invariance (Zueva 2015). Chaotic behavior of healthy brain areas, single neurons, and neural networks has been reported in many works (Babloyantz 1989; Schiff et al. 1994; Faure and Korn 2001; Korn and Faure 2003; Izhikevich 2007).

Multifractal behavior is also noticed in different physiologic processes. Instead of one scaling parameter, multifractal systems are characterized by a singularity spectrum which defines distribution of scaling parameters termed Hölder exponents (a generalization of the Hurst exponent). From the statistical properties of a time series, the Hölder exponent (h) can be interpreted, where a value of $0 \leq h \leq 0.5$ is an indication of anti-persistent behavior, i.e., increment at a particular interval will be followed by a decrement and vice versa. If the value of h corresponds to 0.5, it is an uncorrelated random walk where increase and decrease are equally likely. Finally if the value of $h > 0.5$, persistence is noticed in the time series, i.e., an increment in one particular interval will be followed by a further increment in the next interval (Bishop et al. 2012).

Multifractal analysis is a useful analysis technique for the description of normalized and stationary signals (Stanley and Meakin 1988). It presents observations on the scaling behaviors of time series and also identifies spatial heterogeneity of theoretical and experimental fractal patterns (Grassberger and Procaccia 1983). Fixed-size box-counting algorithm (Halsey et al. 1986) is the most familiar numerical implementation of multifractal analysis. The simplest type of multifractal analysis is based upon the standard partition function multifractal formalism, developed for the multifractal characterization of normalized, stationary measures (Feder 1988; Barabasi and Vicsek 1991; Peitgen et al. 1992; Bacry et al. 2001). For nonstationary time series that are affected by trends or that cannot be normalized, the standard formalism is inefficient in providing correct results. Thus an improved multifractal formalism method was developed, the wavelet transform modulus maxima (WTMM) method (Muzy et al. 1991, 1994; Arneodo et al. 1995, 2000, 2002; Ivanov et al. 1999; Amaral et al. 2001; Silchenko and Hu 2001), which is based on wavelet analysis and involves tracing the maxima lines in the continuous wavelet transform over all scales (Kantelhardt et al. 2002).

1.5.1.1 Detrended Fractal Analysis

External influences, such as seasonal impacts on urban water consumptions (Gato et al. 2007) and business cycles of trade volumes (Barsky and Miron 1989), conceal the correlations in time series. Because of these external influences, correlated data may seem to be uncorrelated, and uncorrelated data with long-term trends may appear correlated. Thus removing these trends effectively is important for identifying the underlying correlations in time series. Rescaled analysis (R/S) (Hurst 1951) along with other non-detrended methodologies (Holschneider 1995) performs better with trend-eliminated long data series. But these methods are not suitable when the time series is nonstationary (Ge and Leung 2013).

Peng et al. (1994) proposed the detrended fluctuation analysis (DFA) method as a fractal scaling method which can detect long-range correlation in nonstationary signals. It quantifies the complexity of signals using the fractal property (Peng et al. 1995; Pikkujamsa et al. 1999). This method is a modified root mean square method for the random walk. Mean square distance of the signal from the local trend line is analyzed as a function of scale parameter. Power-law dependence (Golińska 2012) is usually found, and the exponent α quantifies the degree of correlation in a signal. DFA has successfully been applied to diverse fields such as DNA sequences (Ossadnik et al. 1994; Buldyrev et al. 1995), heart rate dynamics (Peng et al. 1995; Bunde et al. 2000), neuron spiking (Blesic et al. 1999; Bahar et al. 2001), human gait (Hausdorff et al. 1997), long-time weather records (Koscielny-Bunde et al. 1998; Talkner and Weber 2000), cloud structure (Ivanova et al. 2000), geology (Malamud and Turcotte 1999), ethnology (Alados and Huffman 2000), economic time series (Liu et al. 1999; Vandewalle et al. 1999a, b; Mantegna and Stanley 2000), and solid-state physics (Kantelhardt et al. 1999; Vandewalle et al. 1999a, b). The DFA method can avoid spurious detection of correlations that are artifacts of nonstationarities in the time series (Kantelhardt et al. 2002), but cannot characterize time series having multifractal properties. It assumes that both stationary and nonstationary time series are monofractal which can be quantified by a single scaling exponent. However there are some time series which consists of many intermixed fractal subgroups which exhibit multifractal scaling property. By analyzing series of different moments, multifractality can be disclosed. Thus advanced fractal methods have become a necessary requirement for identifying long-range correlation and multifractal properties of time series (Ge and Leung 2013).

Kantelhardt et al. (2002) proposed the multifractal detrended fluctuation analysis (MF-DFA) to determine long-range correlation and multifractal property of time series. MF-DFA has been used to reveal long-range dependency and multifractality in hydrology (Kantelhardt et al. 2003), earthquake (Telesca et al. 2005, Telesca and Lapenna 2006), economics (Du and Ning 2008), magnetic field data (Anh et al. 2007), and sunspot activities (Movahed et al. 2006). Due to the robust nature of this method, it has also found potential use in studying neurological disorders too. In MF-DFA method, a fluctuation function is defined which analyzes multifractal features for different mathematical moments. The aim of the MF-DFA procedure is principally to determine the behavior of the q-dependent fluctuation functions $F_q(s)$ with regard to the time scale s, for various values of order q. Each time series is first transformed according to the MF-DFA algorithm (described in Appendix A), and then some parameters are determined which substantiate the series' multifractality. After obtaining the fluctuation function $F_q(s)$ for different values of order $q = (-10 \text{ to } +10)$, the scaling behavior of the fluctuation function is determined by analyzing the log–log plots of $F_q(s)$ versus log s for each value of q. If the time series are long-range power-law correlated, $F_q(s)$ increases, for large values of s, as a power law, $[F_q(s) \sim s^{h(q)}]$. The linear fit of log $F_q(s)$ versus log s gives the values of the generalized Hurst exponent $h(q)$ which is used to determine the scaling behavior of a time series. For a monofractal time series, $h(q)$ depends on q, and for a multifractal one, there is significant dependence of $h(q)$ on q. Next the classical scaling exponent $\tau(q)$ is evaluated, and a non-linear dependence of $\tau(q)$ on q gives evidence of multifractality in the time series as a linear dependence is indicative of monofractal behavior. To obtain the multifractal spectrum, the singularity strength or Hölder exponent α and dimension of the subset series $f(\alpha)$ are calculated. From the plot of $f(\alpha)$ versus α, the width of the multifractal spectrum (w) is obtained which is believed to be the measure of degree of complexity of a signal. Finally the auto-correlation exponent (γ) is computed which measures how much the signal under consideration is correlated in itself where the lower the value of γ, the higher is the degree of correlation. The method is detailed in Appendix A.

Multifractal analysis using moving average was proposed to determine Hurst exponent of self-affine signals (Vandewalle and Ausloos 1998). The detrending moving average (DMA) method was later developed considering second-order difference between original signal and moving average function (Alessio et al. 2002). The DMA methodology was extended to multifractal detrending moving average (MF-DMA) to analyze multifractal time series and surfaces (Gu and Zhou 2010).

Several variables exist in nature which occurs simultaneously, and they show long-range dependence or multifractal nature. For investigating the long-range cross-correlation between two nonstationary time series, Podobnik and Stanley (2008) proposed detrended cross-correlation analysis (DXA) which is the generalization of DFA method. To determine the correlations between positive and negative fluctuations in a single time series, Jun et al. (2006) introduced detrended cross-correlation method. Podobnik et al. (2009a) used DXA method to reveal long-range power-law cross-correlations in the random part of the underlying stochastic process and the cross-correlation between volume change and price change (Podobnik et al. 2009b). The DXA cross-correlation coefficient (Zebende 2011) calculated for climatological time series (Vassoler and Zebende 2012) is based on overlapping windows. Later Podobnik et al. (2011) calculated cross-correlation coefficient based on nonoverlapping windows and determined its statistical significance both for nonoverlapping and overlapping windows. To determine power-law cross-correlations between different simultaneously recorded time series in the presence of nonstationary sinusoidal and polynomial overlying trends, Horvatic et al. (2011) used DXA with varying orders of polynomial ℓ. This new method was called DXA – $\ell(n)$ (n denoting the scale) was applied to study meteorological data (Dutta et al. 2016).

With a motive to unveil the multifractal features of two cross-correlated signals, Zhou (2008) introduced multifractal detrended cross-correlation analysis (MF-DXA) which is the generalization of DXA method. Wang et al. (2010) investigated the cross-correlations between Chinese A-share and B-share markets using the MF-DXA method and founded the existence of multifractality. He and Chen (2011a, b) used MF-DXA to state that multifractal cross-correlation characteristics are important both in Chinese and US agricultural future markets. Yuan et al. (2012) reported multifractality of cross-correlation between the Chinese stock price and trading volume. Relevance of the method has also been noted using one- and two-dimensional binomial measures, multifractal random walks (MRWs), and financial prices (Podobnik and Stanley 2008; Zhou 2008). The method has also been applied to various physiological time series with high degree of success. We have used the MF-DXA method to quantify the degree of cross-correlation of human gait rhythm in healthy subjects and diseased group and also used it to detect the cross-correlation between ECG and blood pressure signals during change of posture. In MF-DXA, two time series are considered unlikely in MF-DFA where only a single time series is considered. Firstly for two time series in consideration, the fluctuation function F_q is determined. Likewise in MF-DFA here, two scaling behaviors of the fluctuation functions are determined by analyzing log–log plots of fluctuation function F_q versus time scale s. For two cross-correlated time series, a power-law relationship exists between the fluctuation functions and the scale s. The degree of cross-correlation between the two series is given by exponent λ which is obtained from the linear fit of log–log plots of fluctuation function and time scale. The cross-correlation scaling exponent λ is generally average of the Hurst exponent of the two time series. The dependence of scaling exponent (λ) on different orders of q is an indication that the cross-correlated series are multifractal in nature. In this method, multifractal width of the cross-correlated signal (w_x) and the cross-correlation exponent (γ_x) is also evaluated which measures the degree of multifractality and degree of cross-correlation in the considered time series, respectively. A detailed description of the method can be found in Appendix B.

1.5.2 Non-linear Analysis of Biomedical Signals

The fundamental features of biomedical signals may be apparently visible but cannot be described by conventional methods like the average amplitude of the signal. Biomedical signals possess a scale-invariant structure. Fractal analyses have been used frequently to determine the scale-invariant structure in EEG, ECG, EMG, gait analysis, etc. In inter-spike interval of neuron firing, inter-stride interval of human walking, inter-breath interval of human respiration, and inter-beat intervals of the human heart, the observation of scale-invariant features has helped in distinguishing between healthy and pathological conditions (Ivanov et al. 1999; Peng et al. 2002; Zheng et al. 2005; Hausdorff 2007) and between different types of pathological conditions (Wang et al. 2007). In the branching of the nervous system and lungs (Bassingthwaighte et al. 1990; Abbound et al. 1991; Weibel 1991; Krenz et al. 1992) and bone structure (Parkinson and Fazzalari 1994), scale invariance has also been noted. Different studies have advocated the fact that changes of scale-invariant features of biomedical signals imitate alterations in adaptability of physiological process. Fractal structure may change on successful treatment of pathological conditions leading to improvement of health. Thus fractal methodology is the new tool for prognosis and diagnosis of biomedical signals (Ihlen 2012). In this section a brief review of different biomedical signals in the light of non-linear dynamics has been discussed.

1.5.2.1 Non-linear Analysis of EEG Signals

Since brain is widely accepted (Sanei and Chambers 2007; Thankor and Tong 2009) to be a chaotic dynamic system, the generated EEG signals are also chaotic. An EEG signal is also considered to be chaotic when its amplitude changes randomly with time (Rodrıguez-Bermudez and Garcıa-Laencina 2015). Several studies have demonstrated that EEG signals are extremely non-linear and nonstationary. Yuan et al. (2011) used non-linear features for EEG classification, resulting in which an accuracy of 96.5% was achieved.

Non-linear methods based on the Lyapunov exponent (Faust et al. 2015), higher-order spectra (HOS) (Acharya et al. 2012b), information theory and entropy, and intrinsic mode functions (IMF) are used to detect and extract non-linear features of EEG signals of epileptic patients (Kumar et al. 2010; Fu et al. 2015). Acharya et al. (2009, 2012b) extracted the non-linear HOS features, approximation entropy, and sample entropy from EEG segments. To determine the classification performance among normal, interictal, and ictal EEG signals, different classifiers were used. To decompose EEG signals into groups of intrinsic mode functions (IMFs), Fu et al. (2015) applied the empirical mode decomposition (EMD) method. Different artificial intelligent and machine learning techniques are used also for EEG signal classification such as artificial neural network (ANN) (Minasyan et al. 2010; Nasehi and Pourghassem 2013), support vector machines (SVM) (Meier et al. 2008; Shoeb and Guttag 2010; Alotaiby et al. 2015; Samiee et al. 2015), and k-nearest neighbor (KNN) (Fergus et al. 2016).

1.5.2.2 Non-linear Analysis of ECG Signals

Over the decades analysis of ECG signals has been a major research interest in biomedical signal processing. ECG is the widely used method for the cardiac function assessment. It is extensively available and inexpensive process. Quantitative assessment is helpful as subjective interpretations are full of inconsistencies and are inaccurate. One useful method for quantitative analysis of ECG signals is the fractal method. Pikkujamsa et al. (2001) applied DFA to determine the fractal correlation features of RR interval dynamics and the features of HR dynamics which traditional analysis method fail to detect. The authors argued that increased cardiac vulnerability and greater risk of death of patients with and without heart disease may be indicated by the breakdown of fractal organization into beat-to-beat RR interval. Multiscale entropy can discriminate RR intervals from controls and those with heart disorder such as atrial fibrillation (Costa et al. 2002). Another non-linear method, symbolic dynamics, shows the advantage over deterministic and statistical methods in differentiating between ventricular tachycardia and ventricular fibrillation patients (Wessel et al. 2000). Gierałtowski et al. (2012) proposed a multifractal and multiscale method for the DFA of heart rate variability, exploiting the possibility of adapting the multifractal DFA algorithm in order to provide estimates separately at different scales. This method was recently applied for modeling heart rate variability during sleep and blood pressure variability (Solinski et al. 2016; Castiglioni et al. 2017). Ghosh et al. (2013) employed fractal analysis for comparing ECG signals of control subjects and subjects suffering from intracardiac atrial fibrillation. In case of intracardiac atrial fibrillation, they reported higher values of fractal dimension in comparison to normal ECG signals. This implies increase of ECG complexity for intracardiac atrial fibrillation. D'Addio et al. (2014) did fractal analysis of cardiac patients while resting, stress, and early and late recovery phases of ECG stress test and reported drastic change in values of fractal dimension from resting to stress phase. Applying fractal analysis on ECG signal of people performing Kundalini Yoga and Chi meditation, Bhaduri and Ghosh (2016) found increase in the complexity of cardiac dynamics during meditation.

Some researchers have applied different types of entropy method for analyzing the ECG signal. Joseph et al. (2004) determined the effect of reflexological stimulation on heart rate variability and demonstrated increment of Kolmogorov–Sinei entropy in reflexological stimulation compared to relaxed sitting. This implied that the ECG signal turns more random with reflexology. Kamath (2012) examined Renyi and Shannon entropy for distinguishing normal and ventricular tachycardia or fibrillation (VT–VF) subjects. He found Renyi entropy to exceed Shannon entropy with very high sensitivity, specificity, predictivity, and accuracy.

Baumert et al. (2014) evaluated sample entropy and cross-sample entropy of beat-to-beat variability of heart rate and QT interval during head-up tilt and mental arithmetic stress. Reduced sample entropy of RR intervals and cross-sample entropy during head-up tilt were reported. In the case of mental arithmetic stress, significant reduction in coupling was noticed directed from RR to QT. Namazi and Kulish (2016c) examined the relation between the structures of heart rate and the olfactory stimulus (odorant) and reported complexity of the heart rate to be coupled with the molecular complexity of the odorant. Lower fractal heart rate was found for more structurally complex odorant. Odorant with higher entropy has heart rate of lower approximate entropy. The authors opined that the method can be applied and investigated in the case of patients with heart diseases for rehabilitation purpose.

1.5.2.3 Non-linear Analysis of EMG Signals

Electromyography signals have found application in the areas which include neurological diagnosis, neuromuscular and psychomotor research, sports medicine, prosthetics, rehabilitation, and robot limb control (Bu et al. 2009; Fukuda et al. 2011; Gutiérrez et al. 2012; Baspinar et al. 2013).

Surface EMG signals are assumed to be stationary when they are analyzed using linear methods in time and frequency domain considering smaller intervals of 60–1000 ms (Inbar et al. 1986; Bilodeau et al. 1997; Thongpanja et al. 2013; Chen et al. 2014). According to Venugopal and Ramakrishnan (2014), even if the nonstationary characteristics of SEMG signals can be handled by time–frequency method, it is still believed to originate from linear muscular system. Generation of SEMG signals from complex self-regulating system and its multifractal character is evident from the works of Gupta et al. (1997), Arjunan and Kumar (2007), and Marri et al. (2014).

Estimation of the elbow joint angle with the help of artificial neural networks (ANN) was performed by Suryanarayanan et al. (1995) from EMG signals. Hu et al. (2005) calculated FD from filtered SEMG signals in order to discriminate between forearm supination (FS) and forearm pronation (FP) movements. The study demonstrated the usefulness of FD in capturing different motion patterns of SEMG signals.

Mesin et al. (2009) compared the FD of EMG signals with other muscle fatigue indexes. The study reported that changes in conduction velocity have minimum effect on FD of EMG signal; rather FD depends on the synchronization of motor unit from which they inferred that FD is not an index of peripheral but of central fatigue (Belbasis and Fuss 2018).

Mobasser et al. (2007) estimated the elbow-induced wrist force with a fast orthogonal search (FOS), which is a time domain method for rapid non-linear system identification. Ullah and Jung-Hoon (2009) developed a new mathematical model for elbow joint estimation using EMG signals. Arjunan and Kumar (2010) used fractal measures to distinguish SEMG signals of forearm muscles. Artemiadis and Kyriakopoulos (2011) used a state-space model to determine human arm kinematics from myoelectric activity produced by specific muscle groups of the upper arm and forearm. Vogel et al. (2011) employed support vector machines (SVM) to decode human arm–hand system motion from EMG signals. Clancy et al. (2012) proposed non-linear dynamic models to determine joint torque for elbow. To improve the precision of elbow torque value, second- or third-degree polynomial functions (EMGσ) were added to a non-linear parametric model, and the parameters were evaluated using pseudoinverse and ridge regression method (Aung and Al-Jumaily 2013). To evaluate the commencement of fatigue, a procedure was presented in which SEMG signals of biceps brachii muscles and multifractal technique dynamic contraction were used (Marri and Swaminathan 2015, 2016).

Andrade et al. (2006) used the empirical mode decomposition (EMD) technique for filtering EMG signals that can decompose an EMG signal into a set of intrinsic mode functions (IMFs). Sezgin (2012) used higher-order spectra (HOS) for classifying EMG signals. Several authors have used sample entropy to unveil the complexity of SEMG signals (Cashaback et al. 2013; Lake et al. 2002; Molina-Picó et al. 2011; Ruonala et al. 2014).

Thus the classification of EMG signals is also very important for detecting diseases. Handicapped patients or patients suffering from various neurological disorders like Parkinson's, Huntington's, amyotrophic lateral sclerosis, etc. have very different EMG features compared to healthy ones. These patients have movement disorders and different muscle structures too (Chowdhury et al. 2013). In Chap. 4 we have quantified EMG signals of myopathy and neuropathy and compared the results with those of healthy subjects using a non-linear technique. The method successfully distinguished healthy from the diseased.

1.6 Review of Studies on Neurological Disorders

To anticipate future behavior of a time series, it is important to define a set of rules or recognize patterns from the past. Linear regression models like autoregressive (AR), moving average (MA), and autoregressive moving average (ARMA) are broadly used for forecasting time series (Hoon et al. 1996). But forecasting with these methods does not produce accurate results when the time series are non-linear and their characteristics vary with time. Thus several computing techniques with non-linear methods have been developed to solve the problem of linear approach taking into account non-linearity and uncertainty of time series data (Kim et al. 2013). In this section a brief review of different neurological disorders with linear and non-linear techniques is presented.

1.6.1 Epilepsy

Epilepsy is a frequently encountered neurological disorder which occurs when some of the nerve cells erratically acquire huge amounts of electric impulses for a very short interval. By examining the EEG signals, seizure can be anticipated. Seizure warnings can be given a priori by examining the changes in the brain waves so that alerts can be sent to patients hence avoiding damage to the brain or injury during seizure (Kim et al. 2013).

The brain being a non-linear system, the signals obtained from the brain through EEG also have an irregular and complex structure. Thus EEG data of epilepsy is a complex time series. Due to continuous interaction with external components, EEG signals show changes with time (Andrzejak et al. 2001). Noise is present while measuring EEG data. Owing to the non-linearity, abnormalities, and noise, it is difficult to forecast epilepsy EEG data. Thus, an appropriate and accurate forecasting method is necessary (Kim et al. 2013).

Research and advancements in detecting automatic EEG seizures, quantifying and recognizing them began since 1970. In a preliminary study, Gotman (1982) put forward an automated seizure detection technique based on half-wave decomposition of EEG signals. Gabor et al. (1996) later presented a technique analyzing epileptic seizures using wavelets and self-organizing neural networks. Recent years have seen the surge of development of automated methods that can predict seizure using EEG data. Univariate, bivariate, and multivariate algorithms were introduced to tackle seizure detection and prediction based on the EEG analysis of single or multiple electrodes (Stollberger et al. 2000; Saab and Gotman 2005; Bhavaraju et al. 2006). To detect epileptic seizures or other abnormal episodes in EEG automatically, Li (2002) suggested a method based on multi-resolution analysis. Iasemidis et al. (2003) presented an adaptive seizure prediction algorithm (ASPA) built on the merging of short-term maximum Lyapunov exponents (STLmax) among critical electrodes in the pre-ictal period. For predicting epileptic seizures, Gigola et al. (2004) used wavelet analysis built on the evolution of aggregate energy. Prediction algorithms were also presented by Li and Yao (2005) based on the wavelet transform and fuzzy similarity measurements of EEG data. Based on similarity index, Li and Ouyang (2006) proposed the dynamical similarity measure to predict epileptic seizures using EEG data. Liu et al. (2009) introduced particle filtering, and Zandi et al. (2009) analyzed entropy level corresponding to zero-crossing intervals in scalp EEG and its derivatives for seizure prediction. For extracting correlation dimension from intracranial EEG records, Rabbi et al. (2010) employed non-linear methods to design a fuzzy rule-based system for predicting seizures. Some researchers have also used linear methods such as autoregressive and spectral analysis for forecasting seizure from the EEG data (Rogowski et al. 1981; Salant et al. 1998). Kahn and Gotman (2003) applied continuous wavelet transform (CWT) for seizure detection on intracerebral data.

Siuly et al. (2011) introduced least square support vector machine (LS-SVM) to classify epileptic EEG signals. Shen et al. (2013) introduced a method based on a cascade of wavelet-approximate entropy for feature extraction in the epileptic EEG signal classification. For classification they tested three methods, namely, support vector machine (SVM), k-nearest neighbor (kNN), and radial basis function neural network (RBFNN), to find out the one with the best performance. Siuly and Li (2014) introduced a new algorithm, namely, optimum allocation approach, for feature extraction. After extracting features they were evaluated using multiclass least square support vector machine (MLS-SVM) for classification of epileptic EEG signals.

In spite of availability of a variety of drugs and surgical treatments, seizure is still uncurbed in more than 25% of the patients (Pavel et al. 2017). Most of the seizure does not cause long-lasting injury to the brain, but some patients may die due to cardiac or respiratory complications and sudden unexpected death in epilepsy (SUDEP). In those subjects who sleep alone and have generalized tonic–clonic seizures (GTCS), SUDEP is more in them (Lamberts et al. 2012). Heart rate variability (HRV) analyses may help to identify epileptic seizures. Heart rate changes associated with seizures are more generic and have been well-studied (Eggleston et al. 2014).

1.6.2 Dementia (Alzheimer's Disease)

With gradual dysfunction of the neurons and death of brain cells, the central nervous system develops a group of disorders called dementia. It is a syndrome usually seen in aged people in which attention, memory, executive function, visual–spatial ability, and language decrease (Al-Qazzaz et al. 2014b). Different types of dementia include Alzheimer's disease (AD), Parkinson's disease (PD), dementia with Lewy bodies, Creutzfeldt–Jakob disease, normal pressure hydrocephalus, vascular dementia, and front temporal dementia (DeKosky and Marek 2003; Minguez and Winblad 2010). Among the different types of dementia, AD is the most common (Siuly and Zhang 2016).

Hirata et al. (2005) developed software based on the voxel-based specific region analysis for AD, which can automatically analyze 3D MRI data as a series of segmentation, anatomical standardization, and smoothing using a software and Z-score analysis. Li et al. (2007) employed support vector machine (SVM) to determine the hippocampal volume changes in Alzheimer's. The method is helpful in distinguishing Alzheimer's from control. With the help of linear support vector, Klöppel et al. (2008) introduced a computer-aided diagnosis (CAD) procedure for diagnosis of Alzheimer's from MRI scans. For distinguishing aged healthy subjects with Alzheimer's and mild cognitive impairment (MCI), Colliot et al. (2008) employed an automated segmentation method. Taking into account the common risk factors (Siuly and Zhang 2016), Joshi et al. (2010) applied neural network methods (NN) for Alzheimer's classification. To analyze MRI image of Alzheimer's patient, Hamou et al. (2011) presented a computerized technique based on cluster analysis and decision tree. For detecting dementia MRI and EEG – the clinical biomarker – both can detect the changes that occur in the neurons (Al-Qazzaz et al. 2014b). Detection of dementia in its early stages by studying the EEG signals has been reported by many (Helkala et al. 1991; Claus et al. 1998, 2000; Petrosian et al. 2001; Henderson et al. 2006). In a recent study, Hata et al. (2015) investigated EEG of Alzheimer's patients and determined lagged phase synchronization, a measure which connects the functions of the brain. They found Alzheimer's disease patient to show decreased lagged phase synchronization in delta band among the cortical regions.

1.6.3 Parkinson's Disease

Changes in EEG brain signals have also been noticed in Parkinson's disease. EEG of Parkinson's subjects was found to exhibit non-linear features by Stam et al. (1995). Pezard et al. (2001) reported an increased complexity in Parkinson's group compared to control group which was also confirmed in 2013 by Han et al. (2013) who opined that the rhythm of EEG itself has high degree of complexity. Morales and Kolaczyk (2002) used a wavelet-based multifractal methodology to examine structural differences in mediolateral and anteroposterior sway between center-of-pressure (COP) traces of healthy and Parkinson's patients. Interval inter-spike series (ISIs) obtained from the neurons of basal ganglia for control and Parkinson's subjects in wakeful state was also reported to have non-linear temporal structure (Lim et al. 2010).

A person's walking capability is one of the major components of mobility by which dynamics of gait disorder can be understood. Several authors in their work have advocated gait abnormalities (Jankovic and Kapadia 2001; Schaafsma et al. 2003; Hausdorff 2009; Hove et al. 2012). Gait disorders in PD can be characterized by shortened stride length and reduced stride velocity. When linear and non-linear methods are applied simultaneously, gait variability shows complementary characteristics and its change with age and disease (Hausdorff 2005).

DFA has been used to quantify the structure of stride-to-stride variability in PD (Bartsch et al. 2007; Hausdorff 2009; Hove et al. 2012). Some authors reported the time-dependent organization of Parkinsonian tremor to be more regular (lower approximate entropy, ApEn) in PD patients compared to those with the physiological tremor monitored in the healthy control group (Meigal et al. 2012; Vaillancourt and Newell 2000). Kirchner et al. (2014) quantified stride time variability of PD patients and healthy controls using coefficient of variation (CV), DFA, and adaptive fractal analysis (AFA). Dick and Nozdrachev (2015) showed the wavelet characteristic and the multifractal parameters significantly differ in the tremor of healthy subjects and patients with PD. In another study Dick and Nozdrachev (2016) did a comparative analysis of the wavelet, multifractal, and recurrent features of the involuntary oscillations of the trajectory of the isometric force of the hands of healthy volunteers, patients with Parkinson's disease, and subjects with essential tremor syndrome. For essential tremor, the authors observed a significant enhancement of the wavelet spectrum energy and a decrease of the oscillation complexity which was evident from the occurrence of clear peaks in the power spectra, a decrease in the degree of multifractality, the emergence of a quasiperiodic structure in the recurrence diagrams, an increase in determinism, and a decrease of the entropy of recurrence time density. All these trends were found to be increased for the Parkinsonian tremor data. These characteristics enabled them to quantitatively estimate the degree of deviation of motor function from the healthy case. Afsar et al. (2016) applied three different complexity measures, namely, Shannon, Kullback–Leibler, and Klimontovich's renormalized entropies on gait data of patients with PD disease and healthy controls. For stride time variability of gait, they found renormalized entropy technique recognized subjects with 80% sensitivity, whereas 26.7% and 13.3% sensitivity were observed with Shannon entropy and the Kullback–Leibler relative entropy, respectively.

1.6.4 Huntington's Disease

In another neurological disorder, the Huntington's disease major pathological changes are observed in the basal ganglia, causing neural projection loss in the striatum (caudate nucleus and putamen) (Penney and Young 1993). Bylsma et al. (1994) applied quantitative power spectral analysis (PSA) to frontal, temporal, and occipital EEGs of Huntington's disease patients and healthy control subjects and found abnormal behavior of EEG of HD compared to healthy controls. Merrikh-Bayat (2011) estimated the correlation dimension of PD, HD, and ALS and observed that that the average dimension of Parkinson's and Huntington's diseases is more than the average dimension of healthy control subjects, while the average dimension of ALS is less than it. Danoudis and Iansek (2014) compared the stride length and cadence relationship in HD, PD, and healthy subjects using linear regression analysis. They found the scaling of stride length to be disrupted in participants with HD but not the regulation of cadence. The authors concluded that the results obtained from the study will initiate clinicians to develop productive measures to improve mobility and function in people with HD. Bennasar et al. (2016) used accelerometers to determine the intensity of some of the functional symptoms in HD patients. Authors used state-of-the-art selection procedure to determine the most appropriate feature which can help to differentiate healthy from HD subjects. Warner and Sampalo (2016) proposed disease progression models for HD. The models had a combined effect and can predict continuous marker of HD state as a function of age and cytosine–adenine–guanine (CAG – the genetic factor that drives HD pathology) length.

1.6.5 Motor Neuron Disease (MND)

A rare neurodegenerative disorder in older population which causes damage to motor neurons in the cortex, brain stem, and spinal cord is called motor neuron disease (MND). Its manifestation is found in the upper and lower motor neurons, and the bulbar, limb, and respiratory muscles are found to be affected. From the time of inception of the disorder, a patient survives 3–5 years, while some live longer. Respiratory failure is the prime cause of death (Forsgren et al. 1983). The most common variant of motor neuron disease is amyotrophic lateral sclerosis (ALS). It is also known as Lou Gehrig's disease. Since action of the muscles is controlled by the nervous system, electromyography signals can be used to determine the essential characteristics of the disease in individuals (Fattah et al. 2013). Few works have reported the effect of ALS on EMG signal by analyzing it in time and frequency domains (Lambert and Mulder 1957; Lambert 1969; Behnia and Kelly 1991; Leigh and Al-Chalabi 2000; Kasi 2009). Zhou et al. (2012) in a study argued that the recently developed high-density surface electromyography (HD-SEMG) seems to be a suitable method for capturing fasciculation potentials (FPs) compared to intramuscular EMG. Fattah et al. (2012) analyzed EMG signals recorded from healthy and ALS subjects in time and frequency domain. They presented typical features like auto-correlation, zero-crossing rate, and Fourier transform to diagnose ALS disease. To categorize ALS, k-nearest neighbor classifier was used in a leave-one-out cross validation process. In another study, Fattah et al. (2013) introduced discrete wavelet transform (DWT) features to categorize healthy and ALS patients. Tafhim and Kshirsagar (2014) performed a study of EMG classification of signals recorded from bicep muscles during 25%, 50%, and 75% muscle contraction of normal, myopathy, and neuropathy subjects, respectively, using neural network classifiers. Shen et al. (2015) carried out a review and voxel-wise meta-analysis of works to analyze the activities of brain in healthy group and MND subjects to determine common issues in the studies. The study confined to the use of fMRI signals in motor neuron disease. Their work provided reliable results stating that MND is not only restricted to the motor system rather it is a disorder which involves multiple systems including extra-motor cortex areas. The authors also opined that MND causes cognitive dysfunction and deficiencies in social, emotional, and sensory processes are noticed.

Kehri et al. (2017) proposed neuromuscular disease classification from EMG signals based on different combinations of feature extraction methods and types of classifiers. Combination of wavelet transform (WT) and support vector machine (SVM) improved the classification accuracy than other combinations such as DWT with artificial neural network (ANN), independent component analysis (ICA) with multilayer perceptron neural network (MLPN), principal component analysis (PCA) with ANN, and DWT with probabilistic neural network (PNN). Chorage and Sonone (2017) used DWT to analyze SEMG signal for ALS identification. They also considered time domain parameters, like zero-crossing rate and root mean square, and frequency domain parameters like mean frequency and waveform length. From the time domain, frequency domain, and wavelet domain feature extraction, they compared the threshold values for the different parameters in above mentioned domains to identify the ALS patient with more accuracy. In a recent study, Kiran et al. (2018) to compare EMG signals in ALS and control subjects used tunable Q-factor wavelet transform (TQWT). To derive statistical characteristics like mean absolute deviation, interquartile range, kurtosis, mode, and entropy TQWT disintegrated EMG signal into sub-bands. To compare EMG data of ALS and control subjects, the features extracted were tested on k-nearest neighbor and least squares support vector machine classifiers. From the study the authors concluded that in comparison to other techniques, their method was superior as better classification outcomes were obtained.

The following chapters will present details of the study in the areas mentioned above.

References

Abbound S, Berenfeld O, Sadeh D (1991) Simulation of high- resolution QRS complex using ventricular model with a fractal conduction system. Effects of ischemia on high-frequency QRS potentials. Circ Res 68:1751–1760

Acharya UR, Chua CK, Lim TC, Dorithy, Suri JS (2009) Automatic identification of epileptic EEG signals using nonlinear parameters. J Mech Med Biol 9:539–553

Acharya UR, Filippo Molinari S, Sree V, Chattopadhyay S, Ng KH et al (2012a) Automated diagnosis of epileptic EEG using entropies. Biomed Signal Process Control 7:401–408

Acharya UR, Sree SV, Alvin APC, Yanti R, Suri JS (2012b) Application of non-linear and wavelet based features for the automated identification of epileptic EEG signals. Int J Neural Syst 22:1250002

Addison PS (2002) The illustrated wavelet transform handbook. Institute of Physics Publishing, London

Afsar O, Tirnakli U, Kurths J (2016) Entropy-based complexity measures for gait data of patients with Parkinson's disease. Chaos 26:023115

Aike G, Huiming L (1994) Complexity of the brain and neural dynamics. Sci Technol Rev:4

Alados CL, Huffman MA (2000) Fractal long-range correlations in behavioural sequences of wild chimpanzees: a non-invasive analytical tool for the evaluation of health. Ethnology 106:105–116

Al-Angari HM, Sahakian AV (2007) Use of sample entropy approach to study heart rate variability in obstructive sleep apnea syndrome. IEEE Trans Biomed Eng 54:1900–1904

Alessio E, Carbone A, Castelli G, Frappietro V (2002) Second-order moving average and scaling of stochastic time. Eur Phys J B 27:197–200
Alotaiby TN, Abd El-Samie FE, Alshebeili SA, Aljibreen KH, Alkhanen E (2015) Seizure detection with common spatial pattern and support vector machines. In: 2015 International conference on Information and Communication Technology Research (ICTRC), pp 152–155
Al-Qazzaz NK, Abdulazez IF, Ridha SA (2014a) Simulation recording of an ECG, PCG, and PPG for feature extractions. Al-Khwarizmi Eng J 10:81–91
Al-Qazzaz N, Ali S, Ahmad S, Chellappan A, Islam K et al (2014b) Role of EEG as biomarker in the early detection and classification of dementia. Sci World J 2014:Article ID 906038
Amaral LAN, Ivanov PC, Aoyagi N, Hidaka I, Tomono S et al (2001) Behavioral-independent features of complex heartbeat dynamics. Phys Rev Lett 86:6026–6029
Anderson C, Stolz E, Shamsunder S (1998) Multivariate autoregressive models for classification of spontaneous electroencephalographic signals during mental tasks. IEEE Trans Biomed Eng 45:277–286
Andrade OA, Nasuto S, Kyberd P, Sweeney-Reed CM, van Kanijn FR (2006) EMG signal filtering based on empirical mode decomposition. Biomed Signal Process Control 1:44–55
Andrzejak RG, Lehnertz K, Mormann F, Rieke C, David P et al (2001) Indications of nonlinear deterministic and finite-dimensional structures in time series of brain electrical activity: dependence on recording region and brain state. Phys Rev E 64:061907
Anh V, Yu ZG, Wanliss JA (2007) Analysis of global geomagnetic variability. Nonlinear Process Geophys 14:701–708
Arjunan SP, Kumar DK (2007) Fractal theory based non-linear analysis of SEMG. In: 3rd International conference on Intelligent Sensors, Sensor Networks and Information, Melbourne, Australia, Dec. 3–6, pp 545–548
Arjunan SP, Kumar DK (2010) Decoding subtle forearm flexions using fractal features of surface electromyogram from single and multiple sensors. J Neuroeng Rehabil 7:53
Arneodo A, Bacry E, Graves PV, Muzy JF (1995) Characterizing long-range correlations in DNA sequences from wavelet analysis. Phys Rev Lett 74:3293–3296
Arneodo A, Decoster N, Roux SG (2000) A wavelet-based method for multifractal image analysis. I. Methodology and test applications on isotropic and anisotropic random rough surfaces. Eur Phys J B 15:567–600
Arneodo A, Audit B, Decoster N, Muzy JF, Vaillant C (2002) Wavelet based multifractal formalism: applications to DNA sequences, satellite images of the cloud structure, and stock market data. In: Bunde A, Kropp J, Schellnhuber H-J (eds) The science of disaster: climate disruptions, market crashes and heart attacks. Springer, Berlin, pp 27–102
Artemiadis P, Kyriakopoulos K (2011) A switching regime model for the EMG-based control of a robot arm. IEEE Trans Syst Man Cybern B Cybern 41:53–63
Aubert AE, Beckers F, Seps B (2002) Non-linear dynamics of heart rate variability in athletes: effect of training. Comput Cardiol 29:441–444
Augustine A, Prakash RD, Xavier R, Parassery MC (2016) Review of signal processing techniques for detection of power quality events. Am J Eng Appl Sci 9:364–370
Aung YM, Al-Jumaily A (2013) Estimation of upper limb joint angle using surface EMG signal. Int J Adv Robot Syst 10:369
Ayers S (1997) The application of chaos theory to psychology. Theory Psychol 7:373–398
Babloyantz A (1989) Estimation of correlation dimensions from single and multichannel recordings – a critical view. Brain Dyn 2:122–130
Babloyantz A, Salazar JM, Nicolis C (1985) Evidence of chaotic dynamics of brain activity during the sleep cycle. Phys Lett 111A:152–156
Bacry E, Delour J, Muzy JF (2001) Multifractal random walk. Phys Rev E 64:026103
Bahar S, Kantelhardt JW, Neiman A, Rego HHA, Russell DF et al (2001) Long-range temporal anticorrelations in paddlefish electroreceptors. Europhys Lett 56:454
Barabasi AL, Vicsek T (1991) Multifractality of self-affine fractals. Phys Rev A 44:2730
Barsky RB, Miron JA (1989) The seasonal cycle and the business cycle. J Polit Econ 97(3):503–534

Bartsch R, Plotnik M, Kantelhardt JW, Havlin S, Giladi N, Hausdorff JM (2007) Fluctuation and synchronization of gait intervals and gait force profiles distinguish stages of Parkinson's disease. Physica A: Statistical Mechanics and its Applications 383(2):455–465

Baspinar U, Varol HS, Senyurek VY (2013) Performance Comparison of Artificial Neural Network and Gaussian Mixture Model in Classifying Hand Motions by Using sEMG Signals. Biocybernetics and Biomedical Engineering 33(1):33–45

Bassingthwaighte J, Van Beek J, King R (1990) Fractal branchings: the basis of myocardial flow heterogeneities? Ann N Y Acad Sci 591:392–401

Bassingthwaighte JB, Liebovitch LS, West BJ (1994) Fractal physiology. Oxford University Press, New York, p 364

Baumert M, Czippelova B, Ganesan A, Schmidt M, Zaunseder S, Javorka M (2014) Entropy analysis of RR and QT interval variability during orthostatic and mental stress in healthy subjects. Entropy 16:6384–6393

Behbahani S, Jafarnia Dabanloo N, Motie Nasrabadi AA, Teixeira C, Dourado A (2013) Pre-ictal heart rate variability assessment of epileptic seizures by means of linear and non-linear analyses. Anatol J Cardiol 13:797–803

Behnia M, Kelly J (1991) Role of electromyography in amyotrophic lateral sclerosis. Muscle Nerve 14:1236–1241

Belbasis A, Fuss FK (2018) Muscle performance investigated with a novel smart compression garment based on pressure sensor force myography and its validation against EMG. Front Physiol 9:408

Bennasar M, Hicks Y, Clinch S, Jones P, Rosser A (2016) Huntington's disease assessment using tri axis accelerometers. Comput Sci 96:1193–1201

Benson CC, Partha S, Lajish VL, Kumar R (2017) Fractal analysis of MRI data for the improved characterization of brain tumors. Adv Comput Sci Technol 10:1305–1315

Berger H (1929) Über das Elektrenkephalogramm des Menschen (On the EEG in humans). Archiv fur Psychiatrie Nervenkrankheiten 87:527–570

Berntson GG, Bigger JT Jr, Eckberg DL, Grossman P, Kaufmann PG et al (1997) Heart rate variability: origins, methods, and interpretive caveats. Psychophysiology 34:623–648

Bhaduri A, Ghosh D (2016) Quantitative assessment of heart rate dynamics during meditation: an ECG based study with multi Fractality and visibility graph. Front Physiol 7:44

Bhavaraju NC, Frei MG, Osorio I (2006) Analog seizure detection and performance evaluation. IEEE Trans Biomed Eng 53:238–245

Bianchetti A, Trabucch M (2001) Clinical aspects of Alzheimer's disease. Aging Clin Exp Res 13:221–130

Bilodeau M, Cincera M, Arsenault AB, Gravel D (1997) Normality and stationarity of EMG signals of elbow flexor muscles during ramp and step isometric contractions. J Electromyogr Kinesiol 7:87–96

Bishop SM, Yarham SI, Navapurkar VU, Menon DK, Ercole A (2012) Multifractal analysis of hemodynamic behavior intraoperative instability and its pharmacological manipulation. Anesthesiology 117:810–821

Blesic S, Milosevic S, Stratimirovic D, Ljubisavljevic M (1999) Detrended fluctuation analysis of time series of a firing fusimotor neuron. Physica A 268:275–282

Block A, Von Bloh W, Schellnhuber HJ (1990) Efficient box-counting determination of generalized fractal dimensions. Phys Rev A 42:1869–1874

Bogunovic N, Jovic A (2010) Processing and Analysis of biomedical nonlinear signals by data mining methods. In: IWSSIP 2010_17th international conference on Systems, Signals and Image Processing, pp 276–279

Bronzino JD (2000) The biomedical engineering handbook. A CRC handbook published in Cooperation with IEEE Press, pp 184–185

Brown CT, Witschey WRT (2003) The fractal geometry of ancient Maya settlement. J Archaeol Sci 30:1619–1632

Bu N, Okamoto M, Tsuji T (2009) A hybrid motion classification approach for EMG-based human-robot interfaces using Bayesian and neural networks. IEEE Trans Robot 25:502–511
Buczkowski S, Hildgen P, Cartilier L (1998) Measurements of fractal dimension by box-counting: a critical analysis of data scatter. Physica A 252:23–34
Buldyrev SV, Goldberger AL, Havlin S, Mantegna RN, Matsa ME et al (1995) Long-range correlation properties of coding and noncoding DNA sequences: GenBank analysis. Phys Rev E 51:5084
Bunde A, Havlin S, Kantelhardt JW, Penzel T, Peter JH et al (2000) Correlated and uncorrelated regions in heart-rate fluctuations during sleep. Phys Rev Lett 85:3736–3739
Bylsma FW, Peyser CE, Folstein SE, Folstein MF, Ross C et al (1994) EEG power spectra in Huntington's disease: clinical and neuropsychological correlates. Neuropsychologia 32:137–150
Cashaback JG, Cluff T, Potvin JR (2013) Muscle fatigue and contraction intensity modulates the complexity of surface electromyography. J Electromyogr Kinesiol 23:78–83
Castiglioni P, Lazzeroni D, Brambilla V, Coruzzi P, Faini A (2017) Multifractal multiscale dfa of cardiovascular time series: differences in complex dynamics of systolic blood pressure, diastolic blood pressure and heart rate. in Proceedings of the 2017 39th Annual international conference of the IEEE Engineering in Medicine and Biology Society (EMBC), pp 3477–3480, Jeju Island, South Korea, July 2017
Chen X, Liu A, Peng H, Ward RK (2014) A preliminary study of muscular artifact cancellation in single-channel EEG. Sensors 14:18370–18389
Chorage SS, Sonone AB (2017) DWT based identification of amyotrophic lateral sclerosis using surface EMG signal. Int J Res Eng Appl Manag 3:31–35
Chowdhury RH, Reaz MBI, Ali MABM, Bakar AAA, Chellappan K et al (2013) Surface electromyography signal processing and classification techniques. Sensors 13:12431–12466
Clancy EA, Liu L, Pu L, Moyer DVZ (2012) Identification of constant-posture EMG-torque relationship about the elbow using nonlinear dynamic models. IEEE Trans Biomed Eng 59:205–212
Claus JJ, Ongerboer deVisser BW, Walstra JM, Hijdra A, Verbeeten B Jr, van Gool WA (1998) Quantitative spectral electroencephalography in predicting survival in patients with early Alzheimer disease. Arch Neurol 55:1105–1111
Claus JJ, Ongerboer de Visser BW, Bour LJ et al (2000) Determinants of quantitative spectral electroencephalography in early Alzheimer's disease: cognitive function, regional cerebral bloodflow, and computed tomography. Dement Geriatr Cogn Disord 11:81–89
Colliot O, Chételat G, Chupin M, Desgranges B, Magnin B et al (2008) Discrimination between Alzheimer disease, mild cognitive impairment, and normal aging by using automated segmentation of the hippocampus. Radiology 248:194–201
Conte E, Ware K, Marvulli R, Ianieri G, Megna M et al (2015) Chaos, fractal and recurrence quantification analysis of surface electromyography in muscular dystrophy. World J Neurosci 5:205–257
Costa M, Goldberger AL, Peng CK (2002) Multiscale entropy analysis of complex physiologic time series. Phys Rev Lett 89:068102
D'Addio G, Romano M, Maresca L, Bifulco P, Giallauria F, et al (2014) Fractal behavior of heart rate variability during ECG stress test in cardiac patients. In: 8th Conference of the European Study Group on Cardiovascular Oscillations, ESGCO 2014, pp 155–156
Danoudis M, Iansek R (2014) Gait in Huntington's disease and the stride length-cadence relationship. BMC Neurol 14:161
Darbin O, Adams E, Martino A, Naritoku L, Dees D et al (2013) Non-linear dynamics in parkinsonism. Front Neurol 4:211
DeKosky ST, Marek K (2003) Looking backward to move forward: early detection of neurodegenerative disorders. Science 302(5646):830–834
Devous MD Sr (2002) Functional brain imaging in the dementias: role in early detection, differential diagnosis, and longitudinal studies. Eur J Nucl Med Mol Imaging 29:1685–1696

Dick OE, Nozdrachev AD (2015) Nonlinear dynamics of involuntary shaking of the human hand under motor dysfunction. Hum Physiol 41:156

Dick OE, Nozdrachev AD (2016) Features of parkinsonian and essential tremor of the human Hand[1]. Hum Physiol 42:271–278

Du G, Ning X (2008) Multifractal properties of Chinese stock market in Shanghai. Physica A 387:261–269

Dutta S, Ghosh D, Samanta S (2016) Non linear approach to study the dynamics of neurodegenerative diseases by multifractal Detrended cross-correlation analysis—a quantitative assessment on gait disease. Physica A 448:181–195

Eggleston KS, Olin BD, Fisher RS (2014) Ictal tachycardia: the head-heart connection. Seizure 23:496–505

Eke A, Herman P, Kocsis L, Kozak LR (2002) Fractal characterization of complexity in temporal physiological signals. Physiol Meas 23:R1–R38

Elgandelwar SM, Bairagi VK (2016) Analysis of EEG signals for diagnosis of Alzheimer disease. Int J Sci Eng Res 7:529–532

Enescu B, Ito K, Struzik ZR (2004) Wavelet-based multifractal analysis of real and stimulated time series of earthquakes. Annuals of Disaster Prevention Research Institute, Kyoto University, No. 47B

Farina D, Negro F (2012) Accessing the neural drive to muscle and translation to neurorehabilitation technologies. IEEE Rev Biomed Eng 5:3–14

Farina D, Merletti R, Enoka RM (2004) The extraction of neural strategies from the surface EMG. J Appl Physiol 96:1486–1495

Fattah SA, Iqbal MA, Jumana MA, Sayeed Ud Doulah ABM (2012) Identifying the motor neuron disease in EMG signal using time and frequency domain features with comparison. Signal Image Process Int J 3:99–113

Fattah SA, Sayeed Ud Doulah ABM, Iqbal MA, Shahnaz C, Zhu W-P, et al (2013) Identification of motor neuron disease using wavelet domain features extracted from EMG signal. In: IEEE international symposium on circuits and systems (ISCAS 2013), 19–23 May 2013. Beijing, China

Faure P, Korn H (2001) Is there chaos in the brain? I. Concepts of nonlinear dynamics and methods of investigation. Comptes Rendus de l'Académie des Sciences III 324:773–793

Faust O, Acharya UR, Min L, Sputh B (2010) Automatic identification of epileptic and background EEG signals using frequency domain parameters. Int J Neural Syst 20:159–176

Faust O, Acharya UR, Adeli H, Adeli A (2015) Wavelet-based EEG processing for computer-aided seizure detection and epilepsy diagnosis. Seizure 26:56–64

Feder J (1988) Fractals. Plenum Press, New York

Fergus P, Hussain A, David Hignett D, Al-Jumeily KA-A, Hamdan H (2016) A machine learning system for automated whole-brain seizure detection. Appl Comput Inform 12:70–89

Fisher R, van Emde Boas W, Blume W, Elger C, Genton P et al (2005) Epileptic seizures and epilepsy: definitions proposed by the International League Against Epilepsy (ILAE) and the International Bureau for Epilepsy (IBE). Epilepsia 46:470–472

Flynn AC, Jelinek HF, Smith MC (2005) Heart rate variability analysis: a useful assessment tool for diabetes associated cardiac dysfunction in rural and remote areas. Aust J Rural Health 13:77–82

Forsgren L, Almay BGL, Holmgren G, Wall S (1983) Epidemiology of motor neuron disease in Northern Sweden. Acta Neurol Scand 68:20–29

Fu K, Qu JF, Chai Y, Zou T (2015) Hilbert marginal spectrum analysis for automatic seizure detection in EEG signals. Biomed Signal Process Control 18:179–185

Fujiwara K, Miyajima M, Yamakawa T, Abe E, Suzuki Y et al (2016) Epileptic seizure prediction based on multivariate statistical process control of heart rate variability features. IEEE Trans Biomed Eng 63:1321–1332

Fukuda O, Kim J, Nakai I, Ichikawa Y (2011) EMG control of a pneumatic 5-fingered hand using a Petri net. Artificial Life and Robotics 16(1):90–93

Gabor AJ, Leach RR, Dowla FU (1996) Automated seizure detection using a self-organizing neural network. Electroencephalogr Clin Neurophysiol 99:257–266
Gao J, Sultan H, Hu J, Tung WW (2010) Denoising nonlinear time series by adaptive filtering and wavelet shrinkage: a comparison. IEEE Signal Process Lett 17:237–240
Garrett D, Peterson D, Anderson C, Thaut M (2003) Comparison of linear, nonlinear, and feature selection methods for EEG signal classification. IEEE Trans Neural Syst Rehabil Eng 11:141–144
Gato S, Jayasuriya N, Roberts P (2007) Temperature and rainfall thresholds for base use urban water demand modelling. J Hydrol 337(3–4):364–376
Ge E, Leung Y (2013) Detection of crossover time scales in multifractal detrended fluctuation analysis. J Geogr Syst 15:115–147
Ghosh D, Deb A, Dutta K, Sarkar R, Dutta I et al (2004) Multifractality and multifractal specific heat in fragmentation process in 24Mg-AgBr interaction at 4.5 A GeV. Indian J Phys 78:359–362
Ghosh DC, Chakraborty M, Das T (2013) Fractal approach to identify quantitatively Intracardiac atrial fibrillation from ECG signals. Int J Eng Res Appl 3:129–134
Ghosh D, Dutta S, Chakraborty S, Samanta S (2017) Chaos based quantitative electro-diagnostic marker for diagnosis of myopathy, neuropathy and motor neuron disease. J Neurol Neurosci 8 (S4):226
Gierałtowski J, Żebrowski JJ, Baranowski R (2012) Multiscale multifractal analysis of heart rate variability recordings with a large number of occurrences of arrhythmia. Phys Rev E 85(2)
Gigola S, Ortiz F, D'Attellis CE, Silva W, Kochen S (2004) Prediction of epileptic seizures using accumulated energy in a multiresolution framework. J Neurosci Methods 138:107–111
Goldberger AL (1996) Non-linear dynamics for clinicians: chaos theory, fractals, and complexity at the bedside. Lancet 11:1312–1314
Goldberger AL, Rigney DR, West BJ (1990) Chaos and fractals in human physiology. Sci Am 262:42–49
Goldberger AL, Amaral LAN, Hausdorff JM, Ivanov PC, Peng CK et al (2002) Fractal dynamics in physiology: alterations with disease and aging. Proc Natl Acad Sci U S A 99:2466–2472
Golińska AK (2012) Detrended fluctuation analysis (DFA) in biomedical signal processing: selected examples. Stud Logic Grammar Rhetor 29(42):107–115
Gospodinova E (2014) Graphical methods for nonlinear analysis of ECG signals. Int J Adv Res Comput Sci Softw Eng 4:40–44
Gotman J (1982) Automatic recognition of epileptic seizures in the EEG. Electroencephalogr Clin Neurophoysiol 54:530–540
Grassberger P, Procaccia I (1983) Characterization of strange attractors. Phys Rev Lett 50:346–349
Gu G-F, Zhou W-X (2010) Detrending moving average algorithm for multifractals. Phys Rev E 82:11136
Gupta V, Suryanarayanan S, Reddy NP (1997) Fractal analysis of surface EMG signals from the biceps. Int J Med Inform 45:185–192
Gutiérrez Gutiérrez G, López CB, Navacerrada F, Martínez AM (2012) Use of electromyography in the diagnosis of inflammatory myopathies. ReumatologÃa ClÃnica (English Edition) 8(4):195–200
Halsey TC, Jensen MH, Kadanoff LP, Procaccia I, Shraiman BI (1986) Fractal measures and their singularities: the characterization of strange sets. Phys Rev A 33:1141–1151
Hamou A, Simmons A, Bauer M, Lewden B, Wahlund LO et al (2011) Cluster analysis of MR imaging in Alzheimer‘s disease using decision tree refinement. Int J Artif Intell 6:90–99
Han C-X, Wang J, Yi G-S, Che Y-Q (2013) Investigation of EEG abnormalities in the early stage of Parkinson's disease. Cogn Neurodyn 7:351–359
Hassan AR, Siuly S, Zhang Y (2016) Epileptic seizure detection in EEG signals using tunable-q factor wavelet transform and bootstrap aggregating. Comput Methods Prog Biomed 137:247–259

Hata M, Kazui H, Tanaka T, Ishii R, Canuet L et al (2015) Functional connectivity assessed by resting state EEG correlates with cognitive decline of Alzheimer's disease – an eLORETA study. Clin Neurophysiol 127:1269–1278

Hausdorf F (1919) Dimension und ausseres Mass. Math Ann 79:157–179

Hausdorff JM (2005) Gait variability: methods, modeling and meaning. J Neuroeng Rehabil 2:19–27

Hausdorff JM (2007) Gait dynamics, fractals and falls: finding meaning in the stride-to-stride fluctuations of human walking. Hum Mov Sci 26:555–589

Hausdorff JM (2009) Gait dynamics in parkinson's disease: common and distinct behavior among stride length, gait variability, and fractal-like scaling. Chaos 19:026113

Hausdorff JM, Lertratanakul A, Cudkowicz ME, Peterson AL, Kaliton D et al (1985) Dynamic markers of altered gait rhythm in amyotrophic lateral sclerosis. J Appl Physiol 88:2045–2053

Hausdorff JM, Mitchell SL, Firtion R, Peng CK, Cudkowicz ME et al (1997) Altered fractal dynamics of gait: reduced stride-interval correlations with aging and Huntington's disease. J Appl Physiol 82:262–269

He LY, Chen SP (2011a) Nonlinear bivariate dependency between price and volume relationships in agricultural commodity futures markets: a perspective from multifractal detrended cross-correlation analysis. Physica A 390:297–308

He LY, Chen SP (2011b) Multifractal Detrended Cross-Correlation Analysis of agricultural futures markets. Chaos Solitons Fractals 44:355–361

Helkala E, Laulumaa V, Soikkeli R, Partanen J, Soininen H et al (1991) Slow-wave activity in the spectral analysis of the electroencephalogram is associated with cortical dysfunctions in patients with Alzheimer's disease. Behav Neurosci 105:409–415

Henderson G, Ifeachor E, Hudson N, Goh C, Outram N et al (2006) Development and assessment of methods for detecting dementia using the human electroencephalogram. IEEE Trans Biomed Eng 53:1557–1568

Higuchi T (1988) Approach to an irregular time series on the basis of the fractal theory. Physica D 31:277–283

Hirata Y, Matsuda H, Nemoto K (2005) Voxel-based morphometry to discriminate early Alzheimer's disease from controls. Neurosci Lett 382:269–274

Holschneider M (1995) Wavelets : an analysis tool. Clarendon Press/Oxford University Press, Oxford/New York

Hoon MJLD, Van der Hagen THJJ, Schoonewelle H, van Dam H (1996) Why Yule-Walker should not be used for autoregressive modeling. Ann Nucl Energy 23:1219–1228

Horvatic D, Stanley HE, Podobnik B (2011) Detrended cross-correlation analysis for non-stationary time series with periodic trends. Europhys Lett 94:18007

Hove MJ, Suzuki K, Uchitomi H, Orimo S, Miyake Y (2012) Interactive rhythmic auditory stimulation reinstates natural 1/f timing in gait of parkinson's patients. PLoS One 7:e32600

Hu X, Wang Z-z, Ren X-m (2005) Classification of surface EMG signal with fractal dimension. J Zhejiang Univ Sci 6B(8):844–848

Hug F (2011) Can muscle coordination be precisely studied by surface electromyography. J Electromyogr Kinesiol 21:1–12

Huh K-H, Baik J-S, Yi W-J, Heo M-S, Lee S-S et al (2011) Fractal analysis of mandibular trabecular bone: optimal tile sizes for the tile counting method. Imaging Sci Dent 41:71–78

Humeau A, Chapeau–Blondeau F, Rousseau D, Rousseau P, Trzepizur W et al (2008) Multifractality, sample entropy, and wavelet analyses for age-related changes in the peripheral cardiovascular system: preliminary results. Med Phys 35:717–723

Hurst H (1951) Long term storage capacity of reservoirs. Trans Am Soc Civil Eng 116:770–799

Hyman S, Chisholm D, Kessler R, Patel V, Whiteford HA (2006) Mental disorders in disease control priorities in developing countries. In: Jamison DT, Breman JG, Measham AR, Alleyne G, Claeson M, Evans DB (eds), Disease control priorities in developing countries, pp 605–625

Iasemidis LD, Shiau DS, Chaovalitwongse W, Sackellares JC, Pardolas PM et al (2003) Adaptive epileptic seizure prediction system. IEEE Trans Biomed Eng 50:616–627
Ihlen EAF (2012) Introduction to multifractal detrended fluctuation analysis in Matlab. Front Physiol 3:Article141
Inbar GF, Paiss O, Allin J, Kranz H (1986) Monitoring surface EMG spectral changes by the zero crossing rate. Med Biol Eng Comput 24:10–18
Ivanov PC, Amaral LA, Goldberger AL, Halvin S, Rosenblum MG et al (1999) Multifractality in human heartbeat dynamics. Nature 399:461–465
Ivanov P, Amaral LA, Goldberger S, Halvin M, Rosenblum H et al (2001) From 1/f noise to multifractal cascades in heartbeat dynamics. Chaos 11:641–652
Ivanov P, Chen Z, Hu K, Stanley HE (2004) Multiscale aspects of cardiac control. Physica A 344:685–704
Ivanova K, Ausloos M, Clothiaux EE, Ackerman TP (2000) Break-up of stratus cloud structure predicted from non-Brownian motion liquid water and brightness temperature fluctuations. Europhys Lett 52:40
Izhikevich EM (2007) Dynamical systems in neuroscience. The geometry of excitability and bursting. The MIT Press, Cambridge, MA, p 441
Jankovic J (2008) Parkinson's disease: clinical features and diagnosis. J Neurol Neurosurg Psychiatry 79:368–376
Jankovic J, Kapadia AS (2001) Functional decline in parkinson disease. Arch Neurol 58:1611–1615
Joseph P, Acharya UR, Poo CK, Chee J, Min LC et al (2004) Effect of reflexological stimulation on heart rate variability. ITBM-RBM 25:40–45
Joshi S, Shenoy PD, Vibhudendra Simha GG, Venugopal KR, Patnaik LM (2010) Classification of neuro degenerative disorders based on major risk factors employing machine learning techniques. IACSIT Int J Eng Technol 2:350–355
Jovic A, Bogunovic N (2010) Classification of biological signals based on nonlinear features. In: Melecon 2010–2010 15th IEEE Mediterranean Electrotechnical conference, pp 1340–1345
Jun WC, Oh G, Kim S (2006) Understanding volatility correlation behavior with a magnitude cross-correlation function. Phys Rev E 73:066128
Kahn Y, Gotman J (2003) Wavelet based automatic seizure detection in intracerebral electroencephalogram. Clin Neurophysiol 114:898–908
Kamath C (2012) Entropy-based algorithm to detect life threatening cardiac arrhythmias using raw electrocardiogram signals. Middle East J Sci Res 12:1403–1412
Kandel ER, Squire LR (2000) Neuroscience: breaking down scientific barriers to the study of brain and mind. Science 290:1113–1120
Kantelhardt JW, Berkovits R, Havlin S, Bunde A (1999) Are the phases in the Anderson model long-range correlated? Physica A 266:461–464
Kantelhardt JW, Zschiegner SA, Koscielny-Bunde E, Bunde A, Havlin S et al (2002) Multifractal detrended fluctuation analysis of nonstationary time series. Physica A 316:87–114
Kantelhardt JW, Rybski D, Zschiegner SA, Braun P, Koscielny-Bunde E et al (2003) Multifractality of river runoff and precipitation: comparison of fluctuation analysis and wavelet methods. Physica A 330:240–245
Kasi PK (2009) Characterization of motor unit discharge rate in patients with Amytrophic Lateral Sclerosis (ALS). Worcester Polytechnic Institute, May 2009
Kartz M (1988) Fractals and the analysis of waveforms. Comput Biol Med 18:145–156
Kehri V, Ingle R, Awale R, Oimbe S (2017) Techniques of EMG signal analysis and classification of neuromuscular diseases. In: Iyer B, Nalbalwar S, Pawade R (eds) ICCASP/ICMMD-2016. Advances in intelligent systems research. vol 137, pp 485–491. © 2017- The authors. Published by Atlantis Press
Kim HS, Eykholt R, Salas JD (1999) Nonlinear dynamics, delay times and embedding windows. Physica D 127:48–60

Kim SH, Faloutos C, Yang HJ (2013) Coercively adjusted auto regression model for forecasting in epilepsy EEG. Hindawi Publishing Corporation, Computational and mathematical methods in medicine, 2013, Article ID 545613

Kiran PU, Abhiram N, Taran S, Bajaj V (2018) TQWT based features for classification of ALS and healthy EMG signals. Am J Comput Sci Inf Technol 6:19

Kirchner M, Schubert P, Liebherr M, Haas CT (2014) Detrended fluctuation analysis and adaptive fractal analysis of stride time data in Parkinson's disease: stitching together short gait trials. PLoS One 9:e85787

Klöppel S, Stonnington CM, Chu C, Draganski B, Scahill RI et al (2008) Automatic classification of MR scans in Alzheimer's disease. Brain 131:681–689

Korn H, Faure P (2003) Is there chaos in the brain? II. Experimental evidence and related models. C R Biol 326:787–840

Koscielny-Bunde E, Bunde A, Havlin S, Roman HE, Goldreich Y et al (1998) Indication of a universal persistence law governing atmospheric variability. Phys Rev Lett 81:729

Krenz G, Linehan J, Dawson C (1992) A fractal continuum model of the pulmonary arterial tree. J Appl Physiol 72:2225–2237

Krishna PM, Gadre VM, Desai UB (2003) Multifractals: from modeling to control of broadband network traffic. In: Rangarajan G, Ding M (eds) Processes with long-range correlations, Lecture notes in physics, vol 621. Springer, Berlin/Heidelberg, pp 373–392

Kumar SP, Sriraam N, Benakop PG, Jinaga BC (2010) Entropies based detection of epileptic seizures with artificial neural network classifiers. Expert Syst Appl 37:3284–3291

Lake DE, Richman JS, Griffin MP, Moorman JR (2002) Sample entropy analysis of neonatal heart rate variability. Am J Physiol Regul Integr Comp Physiol 283:R789–R797

Lambert EH (1969) Electromyography in amyotrophic lateral sclerosis. In: Norris FH, Kurland LT (eds) Motor neuron diseases: research in amyotrophic lateral sclerosis and related disorders. Grune and Stratton, New York, pp 135–153

Lambert EH, Mulder DW (1957) Electromyographic studies in amyotrophic lateral sclerosis. Mayo Clin Proc 32:441–446

Lamberts RJ, Thijs RD, Laffan A, Langan Y, Sander JW (2012) Sudden unexpected death in epilepsy: people with nocturnal seizures may be at highest risk. Epilepsia 53:253–257

Lehnertz K (2008) Epilepsy and nonlinear dynamics. J Biol Phys 34:253–266

Leigh PN, Al-Chalabi A (2000) Recent advances in amyotrophic lateral sclerosis. Curr Opin Neurol 13:397–405

Li X (2002) EEG analysis with epileptic seizures using wavelet transform. Department of Automation and Computer-Aided Engineering, Chinese University of Hong Kong, Shatin, Hong Kong, 28 Nov 2002

Li X, Ouyang G (2006) Nonlinear similarity analysis for epileptic seizures prediction. Nonlinear Anal Theory Methods Appl 64:1666–1678

Li X, Yao X (2005) Application of fuzzy similarity to prediction of epileptic seizures using EEG signals. In: Proceedings of the 2nd international conference on Fuzzy Systems and Knowledge Discovery (FSKD '05), 3613, pp 645–652

Li S, Shi F, Pu F, Li X, Jiang T et al (2007) Hippocampal shape analysis of Alzheimer disease based on machine learning methods. Am J Neuroradiol 28:1339–1345

Li S, Liu G, Lin Z (2009) Comparisons of wavelet packet, lifting wavelet and stationary wavelet transform for denoising ECG. In: 2nd IEEE international conference on Computer Science and Information Technology, ICCSIT, pp 491–494

Li Y, Wei HL, Billings SA (2011) Identification of time-varying systems using multi-wavelet basis functions. IEEE Trans Control Syst Technol 19:656–663

Li Y, Luo ML, Li K (2016) A multi-wavelet-based time-varying model identification approach for time-frequency analysis of EEG signals. Neurocomputing 193:106–114

Libenson M (2009) Practical Approach to Electroencephalography. Saunders

Lim J, Sanghera MK, Darbin O, Stewart RM, Jankovic J et al (2010) Nonlinear temporal organization of neuronal discharge in the basal ganglia of Parkinson's disease patients. Exp Neurol 224:542–544

Liu Y, Gopikrishnan P, Cizeau P, Meyer M, Peng CK et al (1999) Statistical properties of the volatility of price fluctuations. Phys Rev E 60:1390

Liu D, Pang Z, Wang Z (2009) Epileptic seizure prediction by a system of particle filter associated with a neural network. EURASIP J Adv Signal Process 2009:638534

Lopes R, Betrouni N (2009) Fractal and multifractal analysis: a review. Med Image Anal 13:634–649

Malamud BD, Turcotte DL (1999) Self-affine time series: measures of weak and strong persistence. J Statist Plann Inference 80:173–196

Malarvili M, Mesbah M (2009) Newborn seizure detection based on heart rate variability. IEEE Trans Biomed Eng 56:2594–2603

Mallat S (2002) A wavelet tour of signal processing, 3rd edn. Amsterdam, Elsevier

Mandelbrot B (1967) Hong long is the coast of Britain? Statistical self-similarity and fractional dimension. Science 156(3775):636–638

Mandelbrot B (1977) Fractals: form, chance, and dimension. W. H. Freeman and Company, San Francisco, p 365

Mandelbrot B (1985) Self-affine fractals and the fractal dimension. Phys Scr 32:257–260

Mandelbrot BB (1995) Negative dimensions and Holders, multifractals and their Holder spectra, and the role of lateral preasymptotics in science. J Fourier Anal Appl Kahane special issue 409–432

Mantegna RN, Stanley HE (2000) An introduction to econophysics. Cambridge University Press, Cambridge

Marri K, Swaminathan R (2015) Identification of onset of fatigue in biceps Brachii muscles using surface EMG and multifractal DMA Algorithm. Biomed Sci Instrum 51:107–114

Marri K, Swaminathan R (2016) Analysis of biceps Brachii muscles in dynamic contraction using sEMG signals and multifractal DMA algorithm. Int J Signal Process Syst 4:79–85

Marri K, Jose J, Karthick PA, Ramakrishnan S (2014) Analysis of fatigue conditions in triceps brachii muscle using sEMG signals and spectral correlation density function. In: International conference on Informatics, Electronics and Vision (ICIEV), Dhaka, May 23–24, pp 1–4

Marsden CD (1982) The mysterious motor function of the basal ganglia: the Robert Wartenberg lecture. Neurology 32:514–539

Meier R, Dittrich H, Schulze-Bonhage A, Aertsen A (2008) Detecting epileptic seizures in long-term human EEG: a new approach to automatic online and real-time detection and classification of polymorphic seizure patterns. J Clin Neurophysiol 25:119–131

Meigal AY, Rissanen SM, Tarvainen MP, Georgiadis SD, Karjalainen PA, Airaksinen O, Kankaanpää M (2012) Linear and nonlinear tremor acceleration characteristics in patients with Parkinson's disease. Physiol Meas 33(3):395–412

Meigal AY, Rissanen SM, Tarvainen MP, Airaksinen O, Kankaanpaa M et al (2013) Non-linear EMG parameters for differential and early diagnostics of Parkinson's disease. Front Neurol 4:135

Merletti R, Farina D (2008) Surface EMG processing: introduction to the special issue. Biomed Signal Process Control 3:115–117

Merrikh-Bayat F (2011) Time series analysis of parkinson's disease, huntington's disease and amyotrophic lateral sclerosis. Procedia Comput Sci 3:210–215

Mesin L, Cescon C, Gazzoni M, Merletti R, Rainoldi A (2009) A bidimensional index for the selective assessment of myoelectric manifestations of peripheral and central muscle fatigue. J Electromyogr Kinesiol 19:851–863

Millan H, Kalauzi A, Cukic M, Biondi R (2010) Nonlinear dynamics of meteorological variables: Multifractality and chaotic invariants in daily records from Pastaza, Ecuador. Theor Appl Climatol 102:75–85

Minasyan GR, Chatten JB, Chatten MJ, Harner RN (2010) Patient-specific early seizure detection from scalp EEG. J Clin Neurophysiol 27:163–178
Minguez C, Winblad B (2010) Biomarkers for Alzheimer's disease and other forms of dementia: clinical needs, limitations and future aspects. Exp Gerontol 45:5–14
Mobasser F, Eklund JM, Hashtrudi-Zaad K (2007) Estimation of elbow-induced wrist force with EMG signals using fast orthogonal search. IEEE Trans Biomed Eng 54:683–693
Molina-Picó A, Cuesta-Frau D, Aboy M, Crespo C, Miró-Martínez P et al (2011) Comparative study of approximate entropy and sample entropy robustness to spikes. Artif Intell Med 53:97–106
Monsifrot J, Carpentier EL, Aoustin Y (2004) Sequential decoding of intramuscular EMG signals via estimation of a Markov model. IEEE Trans Neural Syst Rehabil Eng 22:1030–1038
Morales CJ, Kolaczyk ED (2002) Wavelet-based multifractal analysis of human balance. Ann Biomed Eng 30:588–597
Mormann F, Kreuz T, Andrzejak RG, Peter D, Lehnertz K et al (2003) Epileptic seizures are preceded by a decrease in synchronization. Epilepsy Res 53:173–185
Mormann F, Thomas K, Christoph R, Andrzejak RG, Kraskov A et al (2005) On the predictability of epileptic seizures. Clin Neurophysiol 116:569–587
Movahed MS, Jafari GR, Ghasemi F, Rahvar S, Tabar MRR (2006) Multifractal detrended fluctuation analysis of sunspot time series. J Stat Mech Theory Exp 2006(2):1–17
Muzy JF, Bacry E, Arneodo A (1991) Wavelets and multifractal formalism for singular signals: application to turbulence data. Phys Rev Lett 67:3515–3518
Muzy JF, Bacry E, Arneodo A (1994) The multifractal formalism revisited with wavelets. Int J Bifurcation Chaos 4:245–302
Namazi H, Kulish VV (2015) Fractional diffusion based modelling and prediction of human brain response to external stimuli. Comput Math Methods Med 2015:148534
Namazi H, Kulish VV (2016) Fractal based analysis of the influence of odorants on heart activity. Sci Rep 6:38555
Namazi H, Kulish VV, Wong A (2015) Mathematical modelling and prediction of the effect of chemotherapy on cancer cells. Sci Rep 5:13583
Namazi H, Kulish VV, Akrami A (2016a) The analysis of the influence of fractal structure of stimuli on fractal dynamics in fixational eye movements and EEG signal. Sci Rep 6:26639
Namazi H, Kulish VV, Hussaini J, Delaviz A, Delaviz F et al (2016b) A signal processing based analysis and prediction of seizure onset in patients with epilepsy. Oncotarget 7:342–350
Nasehi S, Pourghassem H (2013) Patient-specific epileptic seizure onset detection algorithm based on spectral features and IPSONN classifier. In: International conference on Communication Systems and Network Technologies, pp 186–190, 2013
Nathan DG, Fontanarosa PB, Wilson JD (2001) Opportunities for medical research in the 21st century. JAMA 285:533–534
Ocak H (2009) Automatic detection of epileptic seizures in EEG using discrete wavelet transform and approximate entropy. Expert Syst Appl 36:2027–2036
Ossadnik SM, Buldyrev SV, Goldberger AL, Havlin S, Mantegna RN et al (1994) Correlation approach to identify coding regions in DNA sequences. Biophys J 67:64–70
Oswiecimka P, Kwapien J, Drozdz S, Rak R (2005) Investigating multifractality of stock market fluctuations using wavelet and detrending fluctuation methods. Acta Phys Pol B 36:2447–2457
Parker TS, Chua LO (1989) Practical numerical algorithms for chaotic systems. Springer, New York, pp 193–194
Parkinson I, Fazzalari N (1994) Cancellous bone structure analysis using image analysis. Australas Phys Eng Sci Med 17:64–70
Patrzalek E (2006) Fractals: Useful Beauty General Introduction to Fractal Geometry. In: General Introduction to Fractal Geometry, pp 1–7, Stan Ackermans Institute, IPO Centre for User-System Interaction, Eindhoven University of Technology

Pavei J, Walz R, Marques JLB (2014) Study of biomarkers for prediction of epileptic seizures using ECG. In: Proceedings CBEB 2014 XXIV Brazilian conference on Biomedical Engineering—CBEB 2014 (Uberlândia), pp 1677–1680

Pavei J, Heinzen RG, Novakova B, Walz R, Serra AJ et al (2017) Early seizure detection based on cardiac autonomic regulation dynamics. Front Physiol 8:765

Peitgen HO, Jurgens H, Saupe D (1992) Chaos and Fractals, Springer, New York (Appendix B)

Peng C-K, Buldyrev SV, Havlin S, Simons M, Stanley HE et al (1994) Mosaic organization of DNA nucleotides. Phys Rev E 49:1685–1689

Peng C-K, Havlin S, Stanley HE, Goldberger AL (1995) Quantification of scaling exponents and crossover phenomena in nonstationary heartbeat time series. Chaos 5:82–87

Peng CK, Mietus JE, Liu Y, Lee C, Hausdorff JM et al (2002) Quantifying fractal dynamics of human respiration: age and gender effects. Ann Biomed Eng 30:683–692

Penney JB, Young AB (1993) Huntington's disease. In: Jankovic J, Tolosa E (eds) Parkinson's disease and movement disorders. Williams & Wilkins, Baltimore, pp 205–216

Petersen RC, Roberts RO, Knopman DS, Boeve BF, Geda YE et al (2009) Mild cognitive impairment: ten years later. Arch Neurol 66:1447

Petrosian A, Prokhorov DV, Lajara-Nanson W, Schiffer RB (2001) Recurrent neural network-based approach for early recognition of Alzheimer's disease in EEG. Clin Neurophysiol 112:1378–1387

Pezard L, Jech R, Ruzicka E (2001) Investigation of non-linear properties of multi- channel EEG in the early stages of Parkinson's disease. Clin Neurophysiol 112:38–45

Phinyomark A, Limsakul C, Phukpattaranont P (2009) A comparative study of wavelet denoising for multifunction myoelectric control. In: International conference on Computer and Automation Engineering, ICCAE, pp 21–25

Pikkujamsa SM, Makikallio TM, Sourannder LB, Raiha IJ, Puukka P et al (1999) Cardiac interbeat interval dynamics from childhood to senescence. Comparison of conventional and new measures based on fractals and chaos theory. Circulation 100:393–399

Pikkujamsa SM, Makikallio TH, Airaksinen KEJ, Huikuri HV (2001) Determinants and interindividual variation of R-R interval dynamics in healthy middle aged subjects. Am J Phys Heart Circ Phys 280:H1400–H1406

Pincus SM (1991) Approximate entropy as a measure of system complexity. In: Proc Natl Acad Sci USA, vol 88, pp 2297–2301

Pincus SM (1995) Approximate entropy ApEn as a complexity measure. Chaos 5:110–117

Podobnik B, Stanley HE (2008) Detrended cross-correlation analysis: a new method for analyzing two nonstationary time series. Phys Rev Lett 100:084102

Podobnik B, Grosse I, Horvati D, Ilic S, Ivanov PC et al (2009a) Quantifying cross-correlations using local and global detrending approaches. Eur Phys J B 71:243–250

Podobnik B, Horvatic D, Petersen AM, Stanley HE (2009b) Cross-correlations between volume change and price change. Proc Natl Acad Sci USA 106:22079–22084

Podobnik B, Jiang Z-Q, Zhou W-X, Stanley HE (2011) Statistical tests for power-law cross-correlated processes. Phys Rev E 84:066118

Polat K, Güne S (2007) Classification of epileptiform EEG using a hybrid system based on decision tree classifier and fast Fourier transform. Appl Math Comput 187:1017–1026

Ponnusamy A, Marques JL, Reuber M (2011) Heart rate variability measures as biomarkers in patients with psychogenic nonepileptic seizures: potential and limitations. Epilepsy Behav 22:685–691

Ponnusamy A, Marques JL, Reuber M (2012) Comparison of heart rate variability parameters during complex partial seizures and psychogenic nonepileptic seizures. Epilepsia 53:1314–1321

Poornachandra S (2008) Wavelet-based denoising using subband dependent threshold for ECG signals. Digital Signal Process 18:49–55

Prince M, Bryce R, Albanese E, Wimo A, Ribeiro W et al (2013) The global prevalence of dementia: a systematic review and meta analysis. Alzheimers Dement 9:63–75

Quintero-Rincon A, Pereyra M, Giano CD, Batatia H, Risk M (2016) A new algorithm for epilepsy seizure onset detection and spread estimation from EEG signals. J Phys Conf Ser 705:012032

Quiroga RQ, Garcia H (2003) Single-trial event-related potentials with wavelet denoising. Clin Neurophysiol 114:376–390

Rabbi AF, Aarabi A, Fazel-Rezai R (2010) Fuzzy rule-based seizure prediction based on correlation dimension changes in intracranial EEG. In: Proceedings of the IEEE Engineering in Medicine and Biology Society conference, pp 3301–3304

Reaz MBI, Hussain MS, Mohd-Yasin F (2006) Techniques of EMG signal analysis: detection, processing, classification and applications. Biological Procedures Online 8:11–35

Rhaman M, Karim AHM, Hasan M, Sultana J (2013) Successive RR interval analysis of PVC with sinus rhythm using fractal dimension, Poincare plot and sample entropy method. Int J Image Graphics Signal Process 2:17–24

Richman JS, Moorman JR (2000) Physiological time-series analysis using approximate entropy and sample entropy. Am J Phys Heart Circ Phys 278:H2039–H2049

Rodrıguez-Bermudez G, Garcıa-Laencina PJ (2015) Analysis of EEG signals using nonlinear dynamics and chaos: a review. Appl Math Inf Sci 9:2309–2321

Rogowski Z, Gath I, Bental E (1981) On the prediction of epileptic seizures. Biol Cybern 42:9–15

Ronghua T, Chizhong H, Siyu F, Suming Z, Jinxiang W et al (2001) Correlation analysis of the cognitive function and changes of BEAM and CT scan in patients with Alzheimer's disease. J Neurol Disord Stroke 8:266–269

Ruonala V, Meigal A, Rissanen SM, Airaksinen O, Kankaanpää M et al (2014) EMG signal morphology and kinematic parameters in essential tremor and Parkinson's disease patients. J Electromyogr Kinesiol 24:300–306

Saab ME, Gotman J (2005) A system to detect the onset of epileptic seizures in scalp EEG. Clin Neurophysiol 116:427–442

Salant Y, Gath I, Henriksen O (1998) Prediction of epileptic seizures from two-channel EEG. Med Biol Eng Comput 36:549–556

Samiee K, Kiranyaz S, Gabbouj M, Saramäki T (2015) Long-term epileptic EEG classification via 2D mapping and textural features. Expert Syst Appl 42:7175–7185

Sanei S, Chambers JA (2007) EEG signal processing. Wiley, New York

Sarkar M, Leong TY (2003) Characterization of medical time series using fuzzy similarity-based fractal dimensions. Artif Intell Med 27:201–222

Schaafsma JD, Giladi N, Balash Y, Bartels AL, Gurevich T et al (2003) Gait dynamics in parkinson's disease: relationship to parkinsonian features, falls and response to levodopa. J Neurol Sci 212:47–53

Schellenberg R, Schwarz A (1993) EEG- and EP-mapping--possible indicators for disturbed information processing in schizophrenia? Prog Neuro-Psychopharmacol Biol Psychiatry 17:595–607

Schiff SJ, Jerger K, Duong DH, Chang T, Spano ML, Ditto WL (1994) Controlling chaos in the brain. Nature 370(6491):615–620

Sezgin N (2012) Analysis of EMG signals in aggressive and normal activities by using higher-order spectra. Sci World J 2012:478952

Shen CP, Chen CC, Hsieh SL, Chen WH, Chen JM et al (2013) High-performance seizure detection system using a wavelet-approximate entropy-fSVM cascade with clinical validation. Clin EEG Neurosci 44:247–256

Shen D, Cul L, Cul B, Fang J, Li D et al (2015) A systematic review and meta-analysis of the functional MRI investigation of motor neuron disease. Front Neurol 6:246

Sheng H, Chen YQ (2011) Multifractional property analysis of human sleep EEG signals. In: Proceedings of the ASME 2011 International Design Engineering Technical Conferences & Computers and Information in Engineering conference, August 28–31, 2011, Washington, DC, USA

Shoeb A, Guttag J (2010) Application of machine learning to epileptic seizure detection. In: Proceedings of the 27th international conference on Machine Learning, Haifa, Israel, 2010

Sian J, Gerlach M, Youdim MBH, Riederer P (1999) Parkinson's disease: a major hypokinetic basal ganglia disorder. J Neural Transm 106:443–476

Silchenko A, Hu CK (2001) Multifractal characterization of stochastic resonance. Phys Rev E 63:041105

Simjanoska M, Gjoreski M, Bogdanova A, Koteska B, Gams M, et al (2018) ECG-derived blood pressure classification using complexity analysis-based machine learning. In: Proceedings of the 11th international joint conference on Biomedical Engineering Systems and Technologies (BIOSTEC 2018) – 5, HEALTHINF, pp 282–292

Singh M, Singh M, Paramjeet (2013) Neuro-degenerative disease diagnosis using human gait: a review. IJITKMI 7:16–20

Siuly S, Li Y (2014) A novel statistical framework for multiclass EEG signal classification. Eng Appl Artif Intell 34:154–167

Siuly S, Zhang Y (2016) Medical big data: neurological diseases diagnosis through medical data analysis. Data Sci Eng 1:54–64

Siuly S, Li Y, Wen P (2011) EEG signal classification based on simple random sampling technique with least square support vector machines. Int J Biomed Eng Technol 7:390–409

Solinski M, Gierałtowski J, Zebrowski J (2016) Modeling heart rate variability including the effect of sleep stages. Chaos 26:023101

Song Y (2011) A review of developments of EEG-based automatic medical support systems for epilepsy diagnosis and seizure detection. J Biomed Sci Eng 4:788–796

Soo Y, Sugi M, Nishino M, Yokoi H, Arai T, et al (2009) Quantitative estimation of muscle fatigue using surface electromyography during static muscle contraction. In: 31st IEEE Engineering in Medicine and Biology Society conference, Minneapolis, MN, Sept 3–6, 1, 2975–2978

Stam CJ (2005) Nonlinear dynamical analysis of EEG and MEG: review of an emerging field. Clin Neurophysiol 116:2266–2301

Stam CJ, Jelles B, Achtereekte HA, Rombouts SA, Slaets JP et al (1995) Investigation of EEG non-linearity in dementia and Parkinson's disease. Electroencephalogr Clin Neurophysiol 95:309–317

Stanley HE, Meakin P (1988) Multifractal phenomena in physics and chemistry. Nature 335:405–409

Stanley HE, Amaral LAN, Goldberger AL, Havlin S, Ivanov PC et al (1999) Statistical physics and physiology: Monofractal and multifractal approaches. Physica A 270:309–324

Stollberger C, Finsterer J, Lutz W, Stoberl C, Kroiss A et al (2000) Multivariate analysis based prediction rule for pulmonary embolism. Thromb Res 97:267–273

Sugavaneswaran L, Umapathy K, Krishnan S (2012) Ambiguity domain-based identification of altered gait pattern in ALS disorder. J Neural Eng 9(4):046004

Suryanarayanan S, Reddy NP, Gupta V (1995) Artificial neural networks for estimation of joint angle from EMG signals. In: Proceedings of 17th international conference of the engineering in Medicine and Biology Society, 1

Tafhim M, Kshirsagar P (2014) A Review on EMG Signal Classification for neurological disorder using neural network. In: International conference on Advances in Engineering & Technology – 2014 (ICAET-2014), pp 21–23

Talebinejad M, Chan ADC, Miri A, Dansereau RM (2009) Fractal analysis of surface electromyography signals: a novel power spectrum- based method. J Electromyogr Kinesiol 19:840–850

Talkner P, Weber RO (2000) Power spectrum and detrended fluctuation analysis: application to daily temperatures. Phys Rev E 62:150

Telesca L, Lapenna V (2006) Measuring multifractality in seismic sequences. Tectonophysics 423:115–123

Telesca L, Lapenna V, Macchiato M (2005) Multifractal fluctuations in earthquake-related geoelectrical signals. New J Phys 7:214

Thankor NV, Tong S (2009) Quantitative EEG analysis methods and clinical applications (Artech House, 2009)

Thongpanja S, Phinyomark A, Quaine F, Laurillau Y, Wongkittisuksa B, et al (2013) Effects of window size and contraction types on the stationarity of biceps brachii muscle EMG signals. In: IEEE 7th International Convention on Rehabilitation Engineering and Assistive Technology, 2013, 44:1–44:4

Tzallas AT, Tsipouras MG, Fotiadis DI (2009) Epileptic seizure detection in EEGs using time-frequency analysis. IEEE Trans Inf Technol Biomed 13:703–710

Ullah K, Jung-Hoon K (2009) A mathematical model for mapping EMG signal to joint torque for the human elbow joint using nonlinear regression. In: 4th International Conference on Autonomous Robots and Agents, ICARA 2009

Vaillancourt DE, Newell KM (2000) The dynamics of resting and postural tremor in Parkinson's disease. Clin Neurophysiol 111(11):2046–2056

Vanage AM, Khade RH, Shinde DB (2012) Classifying five different arrhythmias by analyzing the ECG signals. IJCEM Int J Comput Eng Manag 15:75–80

Vandewalle N, Ausloos M (1998) Crossing of two mobile averages: a method for measuring the roughness exponent. Phys Rev E 58:6832–6834

Vandewalle N, Ausloos M, Boveroux P (1999a) The moving averages demystified. Physica A 269:170–176

Vandewalle N, Ausloos M, Houssa M, Mertens PW, Heyns MM (1999b) Non-Gaussian behavior and anticorrelations in ultrathin gate oxides after soft breakdown. Appl Phys Lett 74:1579–1581

Varon C, Caicedo A, Jansen K, Lagae L, Huffel SV (2014) Detection of epileptic seizures from single lead ECG by means of phase rectified signal averaging. In: 36th Annual international conference of the IEEE Engineering in Medicine and Biology Society, Chicago, pp 3789–3790

Vaseghi VS (1996) Advanced Signal Processing and Digital Noise Reduction. John Wiley, New York

Vassoler RT, Zebende GF (2012) DCCA cross-correlation coefficient apply in time series of air temperature and air relative humidity. Physica A 391:2438–2443

Venugopal G, Ramakrishnan S (2014) Analysis of progressive changes associated with muscle fatigue in dynamic contraction of biceps brachii muscle using surface EMG signals and bispectrum features. Biomed Eng Lett 4:269–276

Venugopal G, Navaneethakrishna M, Ramakrishnan S (2014) Extraction and analysis of multiple time window features associated with muscle fatigue conditions using SEMG signals. Expert Syst Appl 41:2652–2659

Vinik AI, Erbas T, Casellini CM (2013) Diabetic cardiac autonomic neuropathy, inflammation and cardiovascular disease. J Diab Invest 4:4–18

Vogel J, Castellini C, van der Smagt PP (2011) EMG-based teleoperation and manipulation with the DLR LWR-III. In: Proceedings IEEE/RSJ international conference on Intelligent Robots and Systems, 2011, pp 672–678

von Campenhausen S, Bornschein B, Wick R, Botzel K, Sampaio C et al (2005) Prevalence and incidence of Parkinson's disease in Europe. Eur Neuropsychopharmacol 15:473–490

Wang G, Huang H, Xie H, Wang Z, Hu X (2007) Multifractal analysis of ventricular fibrillation and ventricular tachycardia. Med Eng Phys 29:375–379

Wang Y, Wei Y, Wu C (2010) Cross-correlations between Chinese A-share and B-share markets. Physica A 389:5468–5478

Warner JH, Sampalo C (2016) Modeling variability in the progression of Huntington's disease a novel modeling approach applied to structural imaging markers from TRACK-HD. CPT Pharmacometrics Syst Pharmacol 5:437–445

Webber CL Jr, Zbilut JP (1984) Dynamical assessment of physiological systems and states using recurrence plot strategies. J Appl Physiol 76:965–973

Weibel ER (1991) Fractal geometry: a design principle for living organisms. Am J Physiol 261:361–369

Weiner MW, Veitch DP, Aisen PS, Beckett LA, Cairns NJ et al (2012) The Alzheimer's disease neuroimaging initiative: a review of papers published since its inception. Alzheimers Dement 8: S1–S68

Wessel N, Ziehmann C, Kurths J, Meyerfeldt U, Schirdewan A et al (2000) Short-term forecasting of life-threatening cardiac arrhythmias based on symbolic dynamics and finite- time growth rules. Phys Rev E 61:733–739

Wink AM, Bullmore E, Barnes A, Bernard F, Suckling J (2008) Monofractal and multifractal dynamics of low frequency endogenous brain oscillations in functional MRI. Hum Brain Mapp 29:791–801

Wittchen HU, Jacobi F, Rehm J, Gustavsson A, Svensson M et al (2011) The size and burden of mental disorders and other disorders of the brain in Europe 2010. Eur Neuropsychopharmacol 21:655–679

Yuan Q, Zhou W, Li S, Cai D (2011) Epileptic EEG classification based on extreme learning machine and nonlinear features. Epilepsy Res 96:29–38

Yuan Y, Zhuang X, Liu Z (2012) Price-volume multifractal analysis and its application in Chinese stock markets. Physica A 391:3484–3495

Yunfeng Wu, Sin CN (2010) A PDF-based classification of gait cadence patterns in patients with amyotrophic lateral sclerosis. In: Annual international conference of the IEEE EMBS Buenos Aires, Argentina, pp 1304–1307, 2010

Zandi SA, Dumont GA, Javidan M, Tafreshi R (2009) An entropy-based approach to predict seizures in temporal lobe epilepsy using scalp EEG. Conf Proc IEEE Eng Med Biol Soc 2009:2228–2231

Zebende GF (2011) DCCA cross-correlation coefficient: quantifying level of cross-correlation. Physica A 390:614–618

Zhang ZG, Liu HT, Chan SC, Luk KDK, Hu Y (2010) Time- dependent power spectral density estimation of surface electromyography during isometric muscle contraction: methods and comparisons. J Electromyogr Kinesiol 20:89–101

Zheng Y, Gao JB, Sanchez JC, Principe JC, Okun MS (2005) Multiplicative multifractal modeling and discrimination of human neuronal activity. Phys Lett A 344:253–264

Zhou WX (2008) Multifractal detrended cross-correlation analysis for two nonstationary time series. Phys Rev E 77:066211

Zhou P, Li X, Nezhad FJ, Rymer WZ, Barkhaus PE (2012) Duration of observation required in detecting fasciculation potentials in amyotrophic lateral sclerosis using high-density surface EMG. J Neuroeng Rehabil 9:78

Zhuo SM, Gan JQ, Sepulveda F (2008) Classifying mental tasks based on features of higher-order statistics from EEG signals in brain–computer interface. Inf Sci 178:1629–1640

Zueva MV (2015) Fractality of sensations and the brain health: the theory linking neurodegenerative disorder with distortion of spatial and temporal scale-invariance and fractal complexity of the visible world. Front Aging Neurosci 7:135

Chapter 2
Multifractal Study of EEG Signal of Subjects with Epilepsy and Alzheimer's

Abstract Epilepsy has been identified as a common disorder of central nervous system affecting a huge size of population. This chapter presents a new approach for studying EEG patterns of the human brain in different physiological and pathological states in epileptic patients and normal people with the help of multifractal detrended fluctuation analysis. The chapter also includes a brief discussion about Alzheimer's diseases and its diagnosis techniques. Further multifractal cross-correlation study was also applied on EEG data taken from patients in both stages – during seizure and in seizure-free interval. The chapter ends with a discussion of how this method can be used as a possible biomarker of epilepsy.

2.1 Introduction

Complex dynamical entities all around us can be determined by set of non-linear differential equations. Non-linearity is the principle behind chaotic behavior of the complex systems. Complex systems may comprise of systems found in the environment, in ecology, in biology, or even in finance (Morales-Matamoros et al. 2009). The human brain is an example of an extensively complicated biological entity that continuously transmits information and also processes it so that an individual can perform the required tasks. The neural system which is made up of billions of nerve cells called neurons performs its task by interacting with the neurons in the central nervous system (CNS) and the peripheral neural system. At the cellular level, messages are transmitted and processed by the neurons with the help of action potentials and neural firing (called spikes). These electrical impulses are recorded in the form of electroencephalogram (EEG) from the scalp (Thakor and Tong 2004).

In a normal healthy state, the signals recorded from the brain have a complex and irregular pattern, but in anomalous circumstances like epileptic seizures, high amplitudes of the quasi-periodic waveforms are noticed (Haghighi and Markazi 2017). A seizure is an abrupt rise of electrical activity in the brain which disrupts the behavior and feelings of a person for a small interval of time. When the brain suffers from repeated seizures caused by abnormal communication of a group of neurons, epilepsy is said to be developed (Morales-Matamoros et al. 2009). Based on clinical

D. Ghosh et al., *Multifractals and Chronic Diseases of the Central Nervous System*,
https://doi.org/10.1007/978-981-13-3552-5_2

manifestation, epileptic seizures can be divided into partial or focal and generalized (Tzallas et al. 2007). When a specific part of the brain encounters enormous synchronous electrical discharge, focal epileptic seizure is said to occur, and when the entire brain is affected by the discharge, a generalized epileptic seizure occurs. Both focal and generalized seizures can develop at all ages but it mostly affects younger and older population (Hassan et al. 2016). People often experience significant problems in personal relationships and employment and develop psychological problems as well. Due to inappropriate medical cure, researchers in the domain of biomedical sciences thought to develop methods to forecast the clinical onset of seizures in order to undergo a surgery. To upgrade prediction algorithms and stimulation-based control techniques (Murphy and Patil 2003; Ker et al. 2011; Berenyi et al. 2012; Lin et al. 2013; Carron et al. 2013; Bergey et al. 2015; Salam et al. 2016; Wang et al. 2016), it is necessary to recognize the manner by which a person goes from a normal state to an abnormal epileptic state. This transition has been poorly understood over the years. The dynamic features of various epileptic disorders (Lopes da Silva et al. 2003a) cannot be described by a single mechanism. Different methods have matured to define the dynamics of seizure generation and termination (Lopes da Silva et al. 2003b; Suffczynski et al. 2004; Breakspear 2005; Wendling et al. 2005; Kim et al. 2009; Marten et al. 2009; Goodfellow et al. 2011; Taylor and Baier 2011; Taylor et al. 2014; Jirsa et al. 2014; Milanowski and Suffczynski 2016; Fan et al. 2016). The dynamic reasons for these transitions are attributed to bifurcation, bistability, excitability, and intermittency (Baier et al. 2012).

2.2 Neurological Disorder: Epilepsy, Alzheimer's, and EEG Data

2.2.1 Epilepsy

Epilepsy is one of the most common neurological disorders (Acharya et al. 2012) of the central nervous system. On an average 0.6–0.8% of population around the world, i.e., around 50 million, people globally are affected with epilepsy (Mormann et al. 2007). Depending on the intensity of epilepsy, persistent seizures may cause changes in perception and behavior (Devinsky and Vazquez 1993; Barnes and Paolicchi 2008; Austin et al. 2011), mild degree of convulsions (Bromfield et al. 2006; Vingerhoets 2006), and temporary loss of consciousness (Blumenfeld 2012; Curia et al. 2014). Around one-third of epileptic patients are drug resistant. Though they receive high dose of epileptic medications, they continue to have seizures. A surgical resection of the damaged brain or location of the seizure origin can eradicate or reduce the occurrence of epileptic seizure. A presurgical evaluation from a team of experienced neurologists, neurophysiologists, neuropsychologists, social workers, radiologists, nurses, and epilepsy neurosurgeons is important before a patient is sent for surgery. A patient's previous health records, physical state, social conditions, seizure symptoms and its intensity, and clinical examinations are essential elements of presurgical evaluation. When all presurgical evaluation directs toward a particular conclusion about occurrence of focal seizure, then a surgery of the patient may be done. EEG and brain imaging, axial computerized tomography (cerebral scanner), magnetic resonance scanning, and positron emission tomography (PET) are the vital diagnostic test to detect epilepsy (Morales-Matamoros et al. 2009).

By now we know that electroencephalography (EEG) is a vital tool to detect the non-linear electrical function of the brain's nerve cells, thus making it an important means for assessment and analysis of epilepsy (Rizvi et al. 2013). EEG signals have been observed to include spikes, sharp waves, and spike-and-wave complexes before seizure, in between two seizures and all along seizures (Li et al. 2012). At present manual scanning of EEG recordings is done to detect epileptic activity, which takes several days to complete (Ocak 2009), thus making detection technique time consuming and error prone (Tzallas et al. 2009). Thus it is the need of the day to develop potential and reliable procedures to identify epilepsy from EEG signals (Tzallas et al. 2007; Guo et al. 2011; Fu et al. 2015; Hassan et al. 2016). An electrographic seizure has four phases: (a) the phase prior to seizure is known as pre-ictal, (b) the period of occurrence of seizure is known as ictal, (c) postictal is the period after seizure, and (d) the period between seizure is called interictal (Morales-Matamoros et al. 2009).

Several researchers have proposed automated methods for detecting epileptic activity (Ocak 2009; Lee et al. 2014; Fu et al. 2015) using Fourier spectral analysis for EEG signal extraction assuming EEG signals to be stationary (Polat and Güne 2007). But owing to complex character of the EEG signals, linear analysis techniques fail to capture the minute changes that occur in the signal which is important to understand to help in diagnosis and prognosis of epileptic seizure. Apart from Fourier transforms, wavelet transforms are also employed to detect EEG time series.

The brain is a complex, non-linear system, whereas the Fourier Spectra is a linear system. To eliminate the discrepancy between these two systems, a new technique of non-linear analysis was developed which may be used as an effective tool for highly complex non-linear systems, namely, brain's dynamical system (Easwaramoorthy and Uthayakumar 2010).

In 1998 according to Andreu et al. (1998), there is no linear analysis technique that can anticipate the start of epileptic crisis. A drastic change in EEG from being irregular in case of healthy brain is noticed when an epileptic seizure starts. The electrical activity of EEG recorded during seizure has been found to show violent amplitude fluctuations but regular rhythmic pattern. Loss of complexity due to coordination of neurons population is observed before and during an epileptic seizure (Morales-Matamoros et al. 2009). Due to violent amplitude fluctuations, the degree of complexity of the brain waves is reduced instead of enhancing (Torres 1991; Andreu et al. 1998; Contreras 2007).

Gutiérrez (2001) introduced an algorithm to gather some characteristic information from the electrocorticographic signal so that waves from epileptic patients can be categorized as epileptic waves. The author used the wavelet method and the correlation function to determine the exact location of epileptogenic focus and its extent. He opined that both methods resemble the waves to be epileptic just like an electroencephalographist would do. The conclusion is significant as the neurosurgeon can arrive at decision faster to operate the focus. Sackellares et al. (2002) established that temporal lobe epilepsy can be defined by episodes of uncontrolled electrical discharges. These discharges comprise of coordinated activity of mesial temporal neurons from the hippocampus. Owing to the chaotic and non-linear nature of the epileptic brain, the authors have hypothesized that there is continuous and sudden shift from in and out of the ictal state because epilepsy self-organizes the brain from chaos to order during phase transitions. When spatiotemporal chaos in the brain fails, seizure produces a mechanism for retreating the brain to a chaotic or normal state.

The detrended fluctuation analysis (DFA) method which has been widely used to characterize non-linear time series revealed the long-range power–law correlation in EEG, indicating time scale invariant and fractal structure (Watters 2000; Watters and Martin 2004). Thus fractal geometry is an essential way to determine epilepsy. Unlike the Fourier spectra, the fractal spectra can determine fractal time series in both amplitude and frequency domain by fractal dimension (FD) (Kulish et al. 2006). Esteller et al. (2002) determined the fractal dimension in cortex electroencephalogram (IEEG, ECoG – electrocorticography) using Katz algorithm and found low FD during pre-ictal period, increased FD at beginning of seizure, and decrease in FD while reaching the lowest level of complexity. Thus they concluded that FD measure is a better precision algorithm which can detect minute changes in seizure waves which is difficult for epileptologist. Fractal dimension of human cerebellum in magnetic resonance images of 24 healthy young subjects was measured by Jing et al. (2003) using the box-counting method. Peiris et al. (2005) used Higuchi's algorithm to calculated fractal dimension of EEG. Correlation between FD of EEG and behavioral microsleeps during a visuomotor tracking task was found to be modest. Janjarasjit and Loparo (2009) analyzed self-similar characteristics of ECoG (electrocorticography) data from an epilepsy patient using the wavelet-based fractal method and observed significantly higher value of spectral exponent compared to other states of the brain.

For forecasting seizure from interictal EEG, Andrzejak et al. (2001) employed correlation dimension and mean phase coherence. Using non-linear analysis of similarity between EEG recordings, Quyen et al. (2001) predicted epileptic seizure in real time. Delay vector variance (DVV) technique was used by Gautama et al. (2003) to improve EEG time series identification of epileptic seizure. To identify EEG signal of epileptic patients, Kannathal et al. (2005) distinguished between correlation dimension (CD), largest Lyapunov exponent (LLE), Hurst exponent (H), and entropy. For seizure detection Nigam and Graupe (2004) made use of LAMSTAR artificial neural network method. Guler et al. (2005) used Lyapunov exponents trained with Levenberg–Marquardt algorithm on EEG signals of epileptic patients to compute the classification precision of recurrent neural networks (RNNs). Spatiotemporal analysis method, wavelet feature extraction method in combination with expert model, and wavelet-based similarity analysis method were used by researchers for predicting epileptic seizures (Winterhalder et al. 2006; Subasi 2007; Ouyang et al. 2007; He et al. 2007). Several other works have been reported where EEG signals have been analyzed in different physiological states (during sleep, awake, through meditation state) of healthy persons and diseased groups including depression, bipolar disorders, Alzheimer's disease, and schizophrenia (Röschke et al. 1995; Fell et al. 2000; Jeongn 2002; Kannathal et al. 2004; Susmakova 2004; Lutz et al. 2004). Among other neurological disorders like Alzheimer's, schizophrenia, and Parkinson's, epilepsy has earned major attention as it the most acquired (Navarro et al. 2002; Winterhalder et al. 2003; D'Alessandro et al. 2005; Schelter et al. 2006; Mormann et al. 2007). If the degree of severity of epileptogenic EEG signals can be understood properly, detection and proper medical cure of epilepsy can be sort for (Ghosh et al. 2014). Litt and Echauz (2002) hypothesized that using methods in time and frequency domain and non-linear and delays methods, different results will be reported. Chaotic assessments such as Lyapunov exponent and fractal dimension (Das et al. 2002); statistical and dynamic non-stationarity analysis (Dikanev et al. 2005); Rényi entropy (Kulish et al. 2006); dynamical similarity index, correlation dimension, and accumulated energy increments (Maiwald et al. 2004); and phase synchronization (Mormann et al. 2003) are some of the other studies conducted to perceive seizure and cultivate a method for prognosis which can be used clinically (López et al. 2009). To classify EEG data of control and epileptic patients, Meghdadi et al. (2008) used a fractal dimension-based technique. Intracranial invasive EEG records reported significantly lower values of correlation fractal dimension in contrast to noninvasive scalp records. Reduced value of correlation dimension was reported for epileptic EEG compared to healthy EEG. Multifractal analysis of the EEG data using the Rényi fractal dimension spectrum too recorded low values and variability in spectrum for seizure activity compared to normal brain activity. Easwaramoorthy and Uthayakumar (2010) designed Advanced form of Generalized Fractal Dimension (GFD) which is a multifractal analysis technique to differentiate between control and ictal EEG. Kamath (2015) examined EEG data from healthy and epilepsy patients using two methods, namely, central tendency measure (CTM) and Higuchi fractal dimension (HFD). CTM quantified degree of variability and HFD complexity. The study suggests that differences exist in the ability to generate random time series between normal and epileptic subjects and between seizure-free and seizure states.

Zhang et al. (2015) analyzed both interictal EEG and ictal EEG using multifractal method. Varied multifractal features differentiated between interictal and ictal phases. Authors also applied relevance vector machine (RVM) for classifying EEG signals. To enhance the procedure's performance on EEG data with epoch-based and event-based estimation techniques, high sensitivity and specificity were achieved. The authors also argued that in event-based performance assessment, a sensitivity of 92.06% with a false detection rate of 0.34/h was obtained.

2.2.2 Alzheimer's Disease

With average life span of human beings being increased over the years, the number of people suffering from Alzheimer's disease (AD) and other forms of dementia has risen considerably. Alzheimer's is a slow and escalating neurodegenerative disorder which causes change in the psychological, behavioral, and functional domain (Ruiz-Gómez et al. 2018). Prevalence of the disease inflates exponentially with age, affecting 1% in the age group of 60 and up to 38% over 85 years (Alzheimer's Association 2017). As Alzheimer's has turned into modern epidemic efforts to explore the underlying brain dynamics has been much researched for. Though current medical care including some therapies are not adequate to treat AD or mild cognitive impairment (MCI), however identification of the problems at an early stage of dementia is significant (Lin and Neumann 2013). Different neuroimaging techniques such as functional magnetic resonance imaging (fMRI), PET, magnetic resonance spectroscopy, EEG, and magnetoencephalography (MEG) (Ewers et al. 2011) are used in differentiating AD subjects from healthy controls. PET and fMRI offer precise results with limited temporal resolution, while EEG and MEG show high temporal resolution, which allows to study the dynamics of complex brain function (Poza et al. 2014). EEG due to low cost, portability and availability is used mostly. Moreover several EEG studies showed its usefulness in characterizing brain dynamics in AD and MCI (Stam 2005; Abásolo et al. 2006; Woon et al. 2007; Gasser et al. 2008; Baker et al. 2008; Hornero et al. 2009; Fernández et al. 2010; Poza et al. 2014). Research using spectral analysis techniques have been traditionally used (Ruiz-Gómez et al. 2018) to detect abnormalities associated with EEG signals obtained from AD and MCI patients (Gasser et al. 2008; Baker et al. 2008). With advent of non-linear analysis techniques, work using these techniques has given information complementary to spectral measures (Stam 2005). Studies suggest more normal EEG activity of patients with AD and MCI in contrast to healthy subjects (Woon et al. 2007; Baker et al. 2008) inferring loss of complexity and indication of disease. Some authors have also reported that with progress in the disease, there is decrease in variability and complexity as well (Abásolo et al. 2006; Hornero et al. 2009; Fernández et al. 2010; Poza et al. 2014). Ruiz-Gómez et al. (2018) in a recent study determined the capability of a method to diagnose EEG using logistic discriminant analysis (LDA), quadratic discriminant analysis (QDA), and multilayer perceptron neural network (MLP). Their methodology was based on both spectral and non-linear features.

Early in 2001, Nagao et al. (2001) introduced a three-dimensional fractal analysis (3D-FA) to determine the spatial heterogeneity of cerebral blood flow (CBF) distribution of single-photon emission computed tomography (SPECT) images in AD. They found comparable difference in values of fractal dimension between control and AD group.

They thus concluded that 3D-FA may be a beneficial tool for objectively determining the advancement of AD. Magnetoencephalogram (MEG) is another noninvasive medical imaging technology which detects the activity of the brain by measuring the magnetic field produced due to electric current flowing within the neurons (Alberdi et al. 2016). Like EEG it is also an important diagnostic tool for measuring brain signals. Gómez et al. (2009a) used MEG signals for diagnosis of AD. Using different methods like sample entropy (SampEn), Lempel-Ziv complexity (LZC) (Gómez et al. 2009a; Gómez and Hornero 2010), Shannon spectral entropy (SSE), approximate entropy (ApEn), Higuchi's fractal dimension (HFD) (Gómez et al. 2009b), Maragos and Sun's fractal dimension (MSFD), and cross-approximate entropy (CrossApEn) (Gómez et al. 2012), different MEG characteristics have been determined which can compare AD and healthy subjects. With the above methods, it was concluded that MEG signals had the ability to distinguish between healthy and diseased subjects with high degree of accuracy. However according to Alberdi et al. (2016) when MCI group was considered, no research has analyzed such high accuracy which could be interesting for predicting AD and ability of MEG signals to detect an initial stage.

Huang-Jing et al. (2015) developed 3D box-counting multifractal analysis (BCMA) and proposed a modified integer ratio-based BCMA (IRBCMA) algorithm too, to study the multifractality of white matter structural changes on 3D MRI volumes between normal aging and early AD. The mutifractal features obtained helped to distinguish AD patients from normal subjects. They opined that IRBCMA algorithm can serve as a more appropriate substitute for 3D volume analysis. The authors recommended the importance of multifractal analysis which can provide explanation about the anatomy of AD by determining the structural changes in white matter (Huang-Jing et al. 2015). Ni et al. (2016) used a multifractal-based approach for studying resting state functional MRI (rs-fMRI) to see whether multifractal features can sufficiently discriminate between healthy and AD and if a combination of multifractal features and other traditional features enhance categorization of AD. Support vector machines and multiple kernel learning (MKL) techniques were used for classification purpose. When multifractal feature (Δf) were combined with monofractal feature using MKL classification, accuracy was found to increase from 69% to 76%. They also opined that with the inclusion of other multifractal features, traditional-feature-based AD classification could be improved.

Since Alzheimer's disease induces functional disconnection between different regions of the brain, some studies have concentrated to determine synchronous changes among pairs of EEG signals (Houmani et al. 2018). To detect the synchronous activity of EEG signals, different methods like correlation coefficient (Dauwels et al. 2010), coherence (Dauwels et al. 2010; Escudero et al. 2011; Sankari et al. 2012), Granger causality (Dauwels et al. 2010; Babiloni et al. 2009), phase synchrony (Dauwels et al. 2010; Czigler et al. 2008; Park et al. 2008), state space-based synchrony (Dauwels et al. 2010; Czigler et al. 2008; Kramer et al. 2007), stochastic event synchrony (Park et al. 2008; Dauwels et al. 2009a, b, 2010; Sankari et al. 2012), and mutual information (Jeong et al. 2001) have been developed. MCI and AD subjects reported reduced EEG synchronies in comparison to healthy controls in all these studies (Houmani et al. 2018). In a recent study, Houmani et al. (2018) performed automated EEG diagnosis on cognitive impairment (SCI) patients, MCI patients, possible AD patients, and patients with vascular dementia, psychosis, Lewy body dementia, and non-neurodegenerative disorders (alcoholism, cerebral vasculitis, cerebellar abscess, etc.). They showed epoch-based entropy (a measure of signal complexity) and bump modeling (a measure of synchrony) can efficiently discriminate among the groups with high accuracy.

All these studies motivated us to perform EEG signal classification using a latest state-of-the-art methodology in the domain of non-linear dynamics. To further extend our understanding of the complex bioelectrical EEG signal, we also performed a cross-correlation study on patients in different states of epilepsy.

2.2.3 EEG Data

The data for the work is obtained from the EEG database made available online by Dr. Ralph Andrzejak et al. (2001) with the Clinic of Epileptology of the University Hospital of Bonn, Germany. The details of the data and data selection procedure are obtained from Andrzejak et al. (2001) [data available online at (www.meb.unibonn.de/epileptologie/science/physik/eegdata.html]. The EEG data consists of five data sets (denoted A–E) each containing 100 single-channel EEG segments of 23.6 s duration. These segments were selected and cut out from continuous multichannel EEG recordings after visual inspection for artifacts, e.g., due to muscle activity or eye movements.

A brief description of the sets is given below:

1. Set A: Extracranial recording of healthy subject with open eyes
2. Set B: Extracranial recording of healthy subject with closed eyes
3. Set C: Intracranial recordings in seizure-free interval from the hippocampal formation of the opposite hemisphere of the brain of patients
4. Set D: Intracranial recordings in seizure-free interval from within epileptogenic zone of patients
5. Set E: Seizure activity of patients

The segments selected here were recorded from positions demonstrating seizure. To record the EEG signals, 128-channel amplifier system was used. The data were written onto the disk of a data acquisition computer system after 12-bit analog-to-digital conversion, at a sampling rate of 173.61 Hz. Band-pass filter settings were 0.53–40 Hz (12 dB/oct) (Andrzejak et al. 2001).

Figures 2.1a, 2.1b, 2.1c, 2.1d, and 2.1e depicts a particular signal from each of the sets (A–E) for 2 s. It is to be noted that we have modified the x-axis of the Figs. 2.1a, 2.1b, 2.1c, 2.1d, and 2.1e without showing the exact time since the information about the amount of time delay is not known to us.

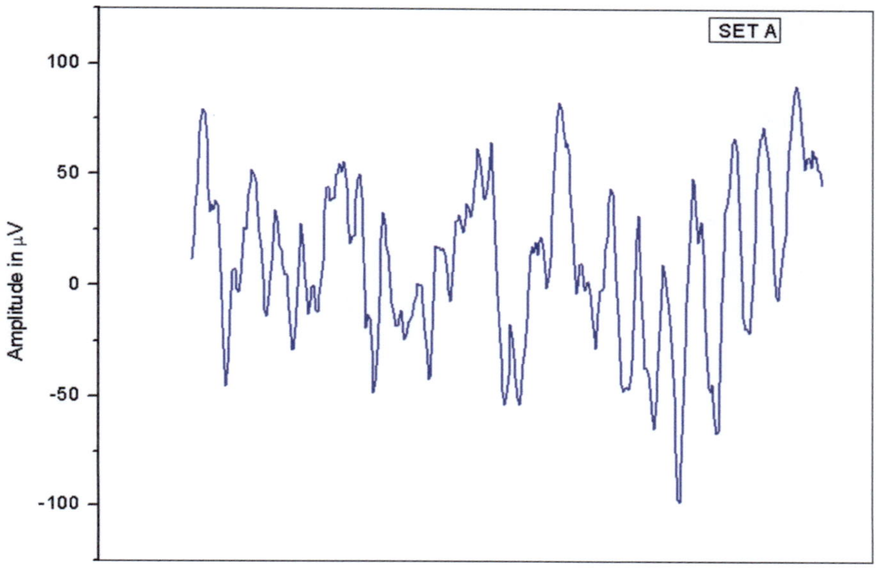

Fig. 2.1a Plot of a typical signal for set A for 2 s

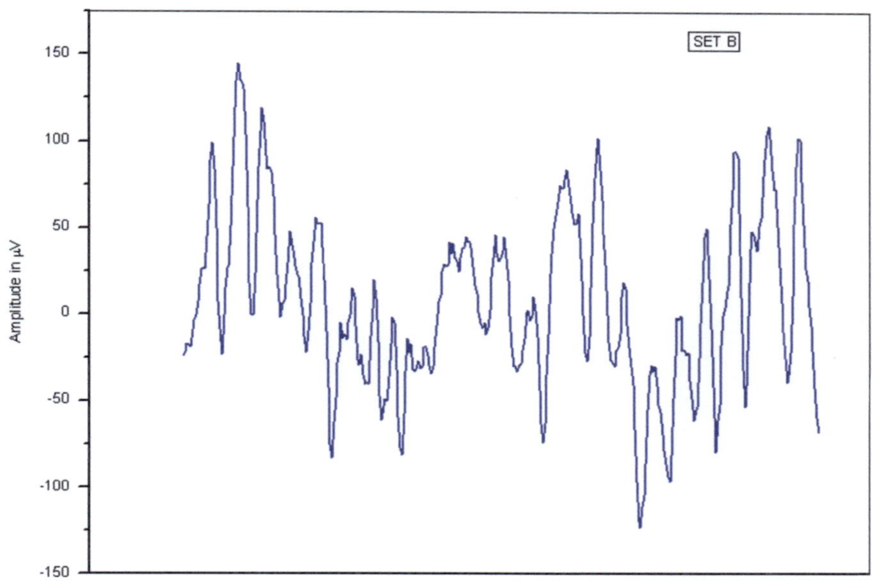

Fig. 2.1b Plot of a typical signal for set B for 2 s

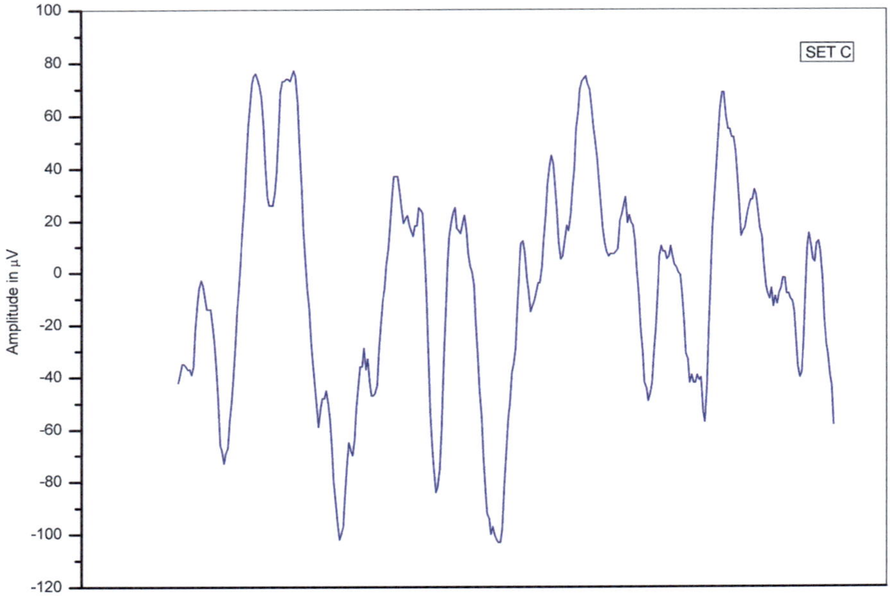

Fig. 2.1c Plot of a typical signal for set C for 2 s

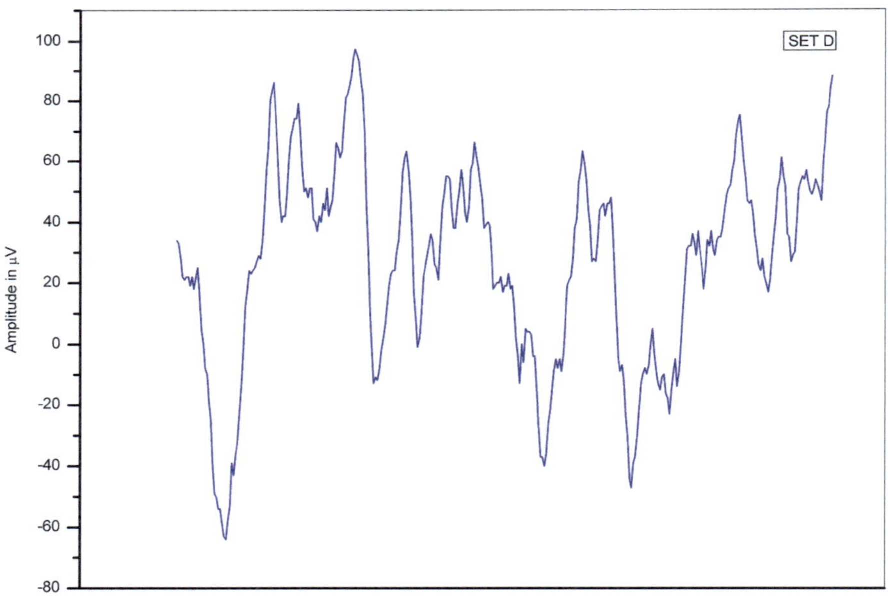

Fig. 2.1d Plot of a typical signal for set D for 2 s

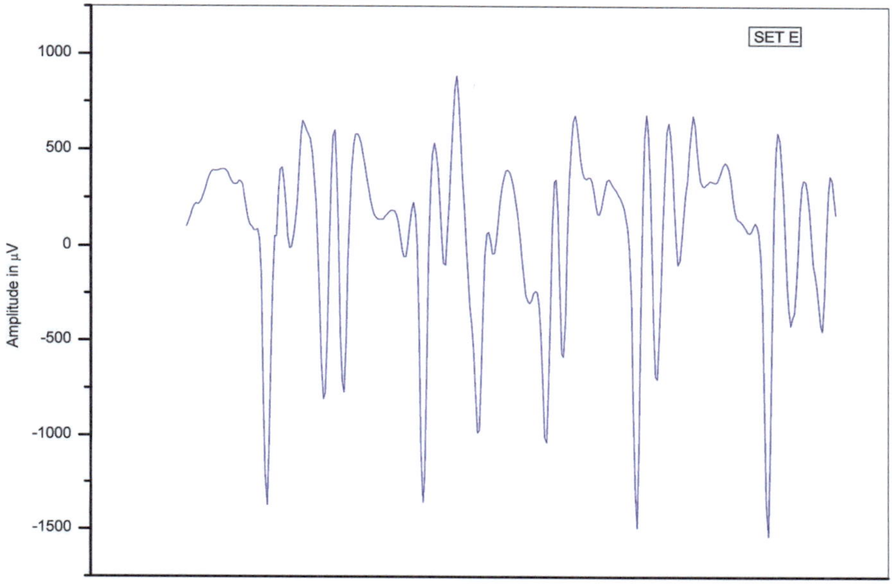

Fig. 2.1e Plot of a typical signal for set E for 2 s

2.3 Multifractal Detrended Fluctuation Analysis of EEG Signals

Due to the spontaneous electrical activity, EEG recorded from the human brain has huge fluctuations. Since the Weierstrass function (Falconer 2003) (everywhere continuous but nowhere differentiable function) characterizes the biomedical waveforms and complex signals, EEG signals are represented as fractal time series (Uthayakumar and Easwaramoorthy 2013). Fractal geometry can be used to analyze EEG from epileptic patients to determine where the epileptogenic region is located, so that a surgery of the epileptic patient can be done (Morales-Matamoros et al. 2009). Osorio and Frei (2007) and Weiss et al. (2008a, b) have also suggested that fractal analysis of intracranial EEG signals can be used in epilepsy research for detection and prediction of focal seizures.

In 2012 Serletis et al. (2012) investigated properties of complexity and multifractality of the activity of neurons in the background during transitions from healthy to epileptic state. The neuronal activities were recorded at the intracellular and local network scales. Reduced complexity and multifractal features were noted for a transition to the epileptic state. Pathology of the degree of multifractality is found to collapse in epileptic state. Thus the authors concluded that the background neuronal activity can adequately acquire complex multifractal features that partially can detect and determine transitions of the brain from healthy to epileptic phase (Serletis et al. 2012). Presence of long-term correlations within the human EEG signals (Parish et al. 2004) has been reported by many investigators which is a fractal characteristic associated with self-similar fluctuations (Nikulin and Brismar 2005). However, neural systems, which interact through feedback of different components, show non-equilibrium, variable fractal behaviors, multifractal (Freeman and Vitiello 2006), and their fractal characteristics can be described by traditional fractal methods, such as detrended fluctuation analysis (DFA) and multiscale entropy (MSE) (Li et al. 2008). In such a situation multifractal analysis is a reliable and relevant method for assessing the fractal characteristics of EEG signals.

Since in traditional non-linear analysis methods like correlation dimension, Lyapunov exponent, approximate entropy, and detrended fluctuation analysis, a single parameter is used, they are insufficient to describe the extremely complex behavior of EEG signals completely. So here we have used a methodology termed multifractal detrended fluctuation analysis (MF-DFA) proposed by Kantelhardt et al. (2002) to study the EEG pattern of normal and epileptic patients. The method has previously been used for analyzing EEG pattern in human and other animals (Figliola 2007; Dojnow 2007; Dutta 2010a).

Using the MF-DFA methodology proposed by Kantelhardt et al. (2002), we have analyzed EEG patterns in healthy and epileptic group. Sets A and B are recordings of normal subjects while sets C, D, and E are EEG records of epileptic patients obtained from an archive of presurgical diagnosis. EEGs of five patients were chosen. All patients had attained complete seizure control after surgery of one of the hippocampal formations which was correctly diagnosed to be the epileptogenic zone. Set C are records of EEG data from hippocampal formation of the opposite hemisphere of the brain of patients in a seizure-free interval while set D corresponds to the EEG data of the epileptogenic zone in a seizure-free interval and set E consists data of seizure activity of patients (Andrzejak et al. 2001).

All the data sets A to E were first changed to obtain the integrated signal according to the MF-DFA algorithm and then the corresponding fluctuation functions $Fq(s)$ were determined. The scaling properties of the fluctuation function for a particular signal for different orders of q are depicted in Fig. 2.2. Linear dependence of the fluctuation function (ln Fq on ln s) is observed indicating scaling behavior. The slope of the linear fit of ln $Fq(s)$ vs. ln s plots gives the values of generalized Hurst exponents $h(q)$. A plot of the Hurst exponent against the order $q(h(q)$ vs. $q)$ is depicted in Fig. 2.3a. It shows variation of $h(q)$ vs. q for a particular set which reveals multifractal behavior as for a monofractal series $h(q)$ is independent of q. We also estimated the classical scaling exponent $\tau(q)$. Variation of $\tau(q)$ with q is shown in Fig. 2.3b. From the plot we observe non-linear dependence of $\tau(q)$ on q indicating multifractal nature of the scaling properties as for monofractal scaling $\tau(q)$ depends linearly on q. Next the singularity strength α and dimension of the subset series $f(\alpha)$ are calculated and their variation known as the multifractal spectrum is shown in Fig. 2.3c (Dutta et al. 2014). By fitting a quadratic function using least square method, the multifractal spectrum can be determined quantitatively (Figliola et al. 2007) around the neighborhood of maximum. Table 2.1 depicts the values of mean multifractal width (w) and variance for each set. The distribution of the values of

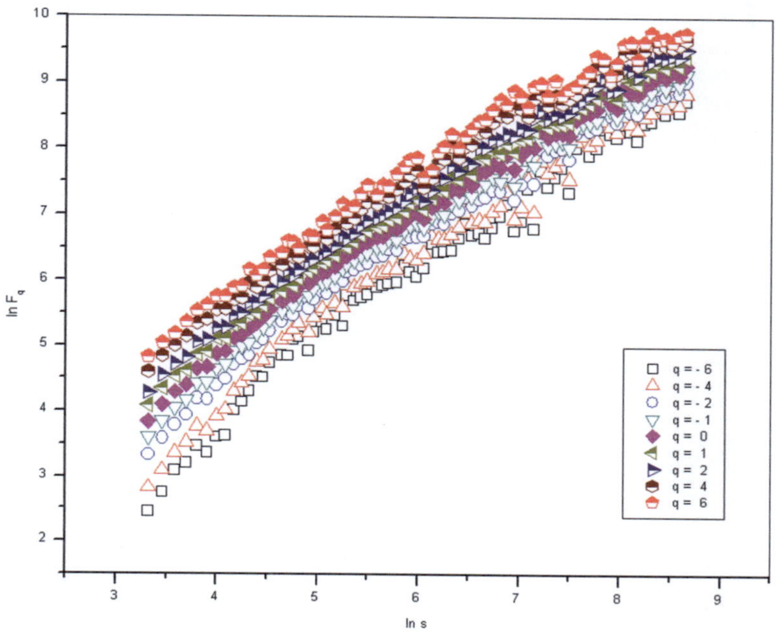

Fig. 2.2 Plot of ln F_q vs. ln s for a particular signal (Dutta et al. 2014)

multifractal width is shown in Fig. 2.4. Since the sample size is quite large, the inference drawn from the results is reasonably significant and the difference in means is also relevant statistically (Dutta et al. 2014). We employed ANOVA (Freund's 2003) to find the statistical significance of results. The means are found to be significantly different even at 99% confidence level.

To determine multifractality origin, we shuffled all the series randomly and then analyzed them following the same procedure as for the original series. In the shuffling procedure, all correlations are disturbed. For the shuffled series too, we estimated values of Hurst exponent $h(q)$, classical scaling exponent $\tau(q)$, and multifractal widths. For comparison with original series, the variation of $h(q)$ vs. q, $\tau(q)$ vs. q, and $f(\alpha)$ vs. α for the shuffled series are also shown in Figs. 2.3a, 2.3b, and 2.3c, respectively. Weaker multifractality is observed for the randomly shuffled series indicating that the origin of multifractality is due to both long-range correlations and broad probability distribution. Ideally for a sufficiently long series, the shuffled series should have monofractal properties with a value of Holder exponent α, close to 0.5. Figure 2.3c shows that for the shuffled series, $f(\alpha)$ vs. α has a peak at α_0 (value of α at $q = 0$) close to 0.5. Ideally $f(\alpha)$ should be independent of α. In this case the series is comparatively short (Dutta et al. 2014). It was reported by Drożdż et al. (2009) that a relatively short series may disclose traces of multifractality; however the results systematically and steadily approach toward monofractal behavior with the increase in number of data points.

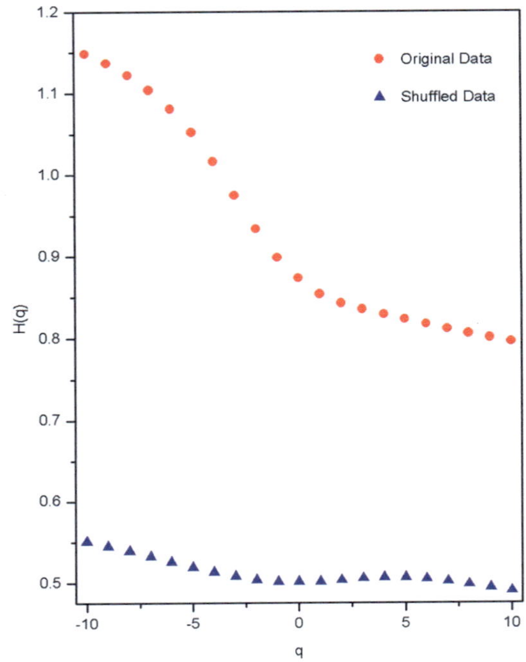

Fig. 2.3a Plot of $H(q)$ vs. q for a particular signal and its corresponding randomly shuffled series (Dutta et al. 2014)

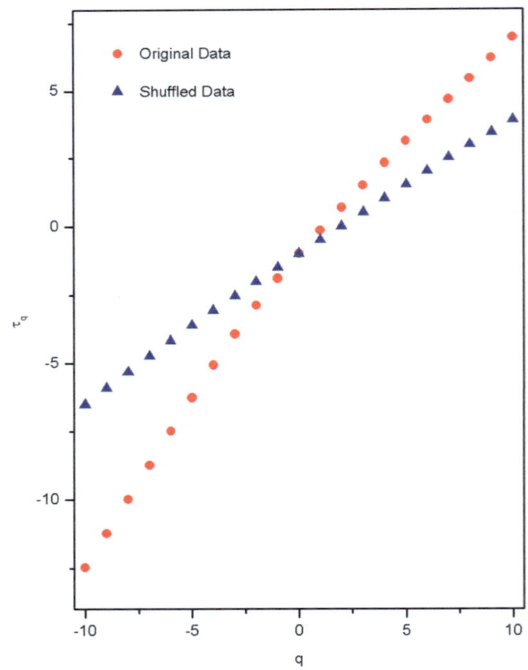

Fig. 2.3b Plot of $\tau(q)$ vs. q for a particular signal and its corresponding randomly shuffled series (Dutta et al. 2014)

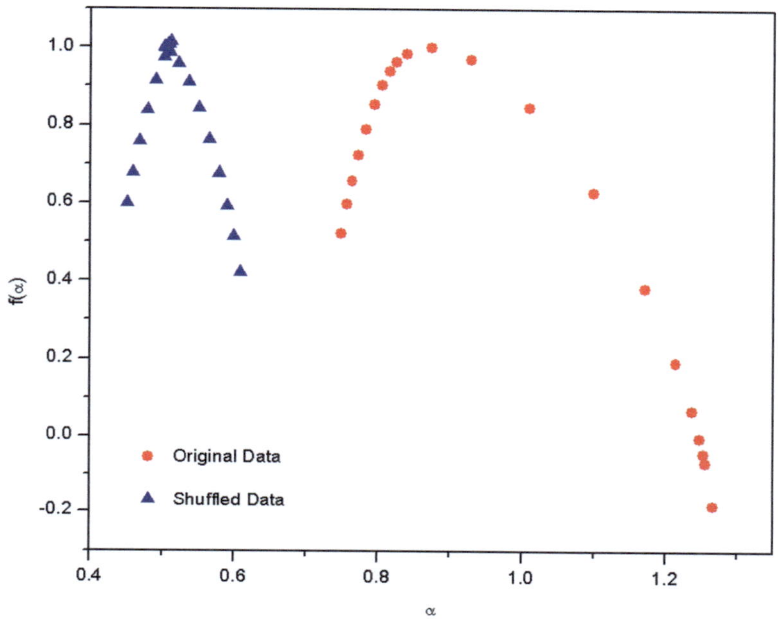

Fig. 2.3c Plot of $f(\alpha)$ vs. α for a particular signal and its corresponding randomly shuffled series (Dutta et al. 2014)

Table 2.1 Mean values of multifractal width and variance for the sets A–E (Dutta et al. 2014)

Set	Mean width	Variance
A	0.605	0.008
B	0.74	0.01
C	0.80	0.01
D	0.92	0.03
E	1.19	0.07

We can draw the following inferences from the values obtained in Table 2.1:

1. The values clearly indicate that the multifractal widths of EEG patterns of the brain are significantly different for extracranial (set A, B) and intracranial recordings (set C, D, E). The results obtained for different physiological states (eyes open and closed) and pathological states (normal and epileptic patients) are also found to be considerably different (Dutta et al. 2014).
2. Multifractal widths of normal patients with eyes open and closed are significantly different. The multifractal width seems to increase with closed eyes (Dutta et al. 2014). As suggested by Andrzejak et al. (2001), the closing of eyes might impose a constraint in the dynamics resulting in physiological alpha rhythm. Alpha waves are reduced with open eyes, drowsiness, and sleep which might lead to the increase of multifractal width for closing of eyes as observed in our case. For

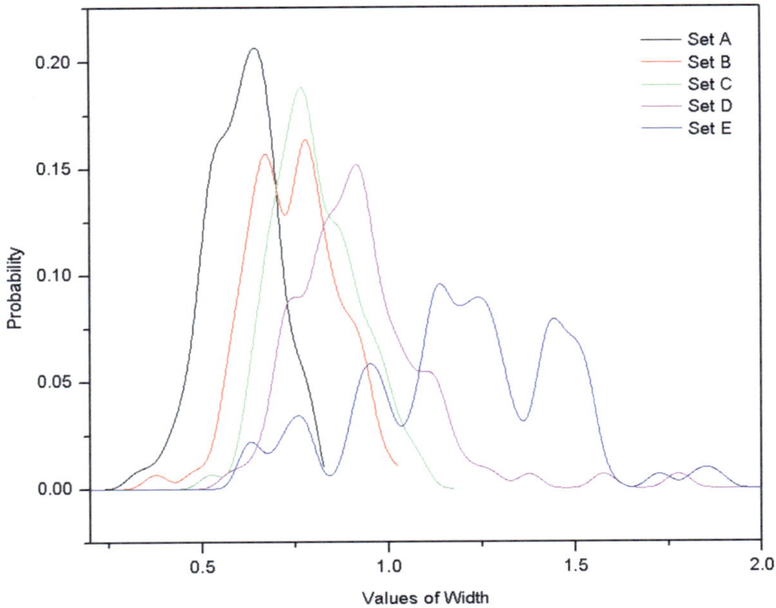

Fig. 2.4 Distribution of values of the multifractal width *W* for sets A–E (Dutta et al. 2014)

closed eyes Stam et al. (1996) have observed low values of correlation dimension D2, compared to eyes open and eyes closed with mental arithmetic. Jelles et al. (1999) have also observed an increase in value of D2 for activation like opening of eyes and arithmetic. In accordance with the previous studies, we have also observed a clear distinction of multifractal parameters for eyes open and closed (Dutta et al. 2014).

3. Now considering the pathological states of the brain, we can see significant difference in the values of multifractal width for normal healthy people and diseased sets (Dutta et al. 2014). Data of set C is obtained from the hippocampal formation on the opposite hemisphere of the brain thus C does not lay in epileptogenic zone. Even though C does not lie in epileptogenic zone, the mean value of multifractal width w is different from that of A or B. Data of sets A and B corresponds to extracranial recording while C to intracranial recording (Dutta et al. 2014) may be the cause of different values of multifractal width. Andrzejak et al. (2001) have explained the possible reasons for the difference in multifractal widths of sets A and C. They have explained that for intracranial recordings fewer neurons contribute to measured potentials and the data are also less filtered compared to the extracranial locations. Further they have also highlighted the fact that though C does not lie directly in the epileptogenic zone, it may participate in secondary non-autonomous processes generated by the epileptogenic zone.

4. Data of set D which is recorded in seizure-free interval from the epileptogenic zone also has a higher degree of multifractality compared to set C. Thus the MFDFA method can be very effective in detection of the epileptogenic zone of the brain. A resection of the epileptogenic zone can result in complete cure of the patient. We finally observe that the EEG recording for seizure activity, i.e., set E, shows the highest degree of multifractality compared to all other sets (Dutta et al. 2014).

It is also worth mentioning that the multifractal width varies "significantly" for different physiological and pathological cases. Compared to the control group, a complex pattern is observed in case of the diseased set. Earlier investigations carried out on cardiac systems as well as EEG patterns of rats (Ivanov et al. 1999, 2001, 2009; Bartsch et al. 2005; Dutta 2010a, b) healthy subjects showed higher degree of multifractality in contrast to our case, though the comparison is more meaningful in similar systems as the underlying processes for heart rate variations and brain functions are not similar. The human brain is a more complex system than the cardiac system, and in an epileptic seizure, the human brain produces huge spikes and shows large fluctuations compared to the normal brain. Since the sample size in the work of Dutta et al. (2010a) is small compared to the present case, hence different results were observed.

Thus the study reveals very interesting results. Bashan et al. (2008) have noticed that MFDFA1 consistently inflates the scaling exponents for small scale, 's'. Another method called centered moving average (CMA) proposed by Alvarez-Ramirez et al. (2005) can be more effective in determining the scaling properties in short time series without trends. However by applying standard DFA method for data with possible unknown trends with several different detrending polynomial orders will be helpful in distinguishing real crossovers and artificial crossovers due to trends. Another problem arises when the correlation properties of real data cannot be identified where a large amount of data is missing or removed due to artifacts. Ma et al. (2010) observed that even extreme loss of data up to 90% does not significantly affect the global scaling behavior of positively correlated signals, but the anticorrelated signals significantly change their scaling behavior even if only 10% of the data are removed. The change is more for a lower value of the scaling exponent. Xu et al. (2011) have also studied how coarse-graining methods affect the scaling properties of long-range correlated and anticorrelated signals which are quantified using the DFA method. Coarse-graining methods strongly affect anticorrelated signals, whereas they have a much weaker effect on the scaling behavior of positively correlated signals. There are also chances of spurious multifractality as discussed by Ludescher et al. (2011). In the case of physiological signals, they have observed the artifacts that create a problem in detecting long-range correlations and hence multifractality is the additive random noise. Additive noise scarcely results in any spurious multifractality in case of monofractal records. Errors can be observed in values of $h(q)$ for $q < 2$ for moderate additions of noise, whereas for $q > 2$ no significant change is observed. But in case of multifractal records, multifractality is highly reduced with the increase in white noise leading to almost monofractal trends. In our study we have obtained large values of the multifractal width for almost all cases which is an indication of strong degree of multifractality. Thus any trends have hardly affected the data. Moreover we have analyzed 100 signals in each case, so the large sample size is expected to give true results when the mean is computed. Epilepsy is a serious neurological disorder which affects a considerable percentage of the human population; therefore the present investigation provides very significant and useful new data which would help in proper diagnosis and treatment of the patient and in complete cure of the patient (Dutta et al. 2014).

2.4 Multifractal Detrended Cross-Correlation Analysis of EEG Signals

With the introduction of multifractal detrended cross-correlation analysis (MF-DXA) by Zhou (2008), the multifractal features of two cross-correlated signals and higher-dimensional multifractal measures could be described. For more than 20 years, several researchers have applied the cross-correlation technique to study EEG as one of more conventional analysis tools as compared to newer tools (Jerger et al. 2001). Cross-correlation was used by Mars and Lopes da Silva (1983) for determining the location of epileptogenic foci. Lopes da Silva et al. (1989) investigated the interdependence of EEG signals, Harris et al. (1994) estimated time delays between channels, and Mann et al. (1993) characterized dynamic properties of sleep EEG using cross-correlation. Zhao et al. (2013) applied DXA to analyze different physiological and pathological states of epilepsy EEG signals and found the DXA values of epilepsy patients' EEG signals increased compared to the normal subjects' EEG signals which can help in medical diagnosis and treatment. Jun and Da-Qing (2012) also used the DXA to study the EEG of healthy young and old subjects and found the cross-correlation between different leads of a healthy young subject was larger than that of a healthy old subject. They also showed the cross-correlation relationship to decrease with age. Bob et al. (2010) found cross-correlations between pairs of EEG time series to be inversely related to dissociative symptoms (psychometric measures) in 58 patients with paranoid schizophrenia (Timasheva et al. 2012, Ghosh et al. 2014).

In our work we studied the cross-correlation of EEG signals in epileptic patients during seizure and in seizure-free interval using MF-DXA methodology. MF-DXA is used with high degree of success in the investigation of complex signals produced by real biological systems. It is a very rigorous and robust tool for assessment of cross-correlation among two non-linear time series – in this case among seizure and seizure-free intervals for epileptic patients (Ghosh et al. 2014). A brief description of the data is given in Sect. 2.2. In the present study, we have performed cross-correlation analysis among sets C, D, and E.

Like in MF-DFA analysis here also the integrated time series was divided into Ns segments [$N_s = \text{int}(N/s)$], N being the length of the series. With increment in the value of q from $= -10$ to $+10$, the fluctuation function for both auto-correlated and cross-correlated series was determined. For all values of q, power–law scaling of fluctuation function with scale s was observed. The slope of linear fit to log $Fxyq(s)$ vs. log s plots gives the values of $\lambda(q)$. The plot of log $Fxq(s)$ and $Fyq(s)$ vs. log s for the auto-correlation (MF-DFA) and log $Fxyq(s)$ for cross-correlation (MF-DXA) fluctuation functions of set C and E and set D and E for $q = 2$ for a particular signal is depicted in Figs. 2.5a and 2.5b. The linear nature of all the curves suggests that there exist power–law cross-correlations. The values of $h(q = 2)$ and $\lambda(q = 2)$ are provided in Tables 2.2 and 2.3, respectively. The relation where cross-correlation scaling exponent $\lambda(q)$ is average of the Hurst exponent $h(q)$ of individual series is observed to be valid in all the cases. The mean values of auto-correlation (γ) are provided in Table 2.2 and the mean values of cross-correlation (γ_x) for the set CE and set DE are presented in Table 2.3 (Ghosh et al. 2014).

Figures 2.6a and 2.6b depicts the relationship between the cross-correlation scaling exponent $\lambda(q)$ and q for each of the sets CE and DE. For comparison (in the same Figs. 2.6a and 2.6b), we have plotted the values of generalized Hurst exponent $h(q)$ of the sets C, D, and E estimated by means of MF-DFA. From the plots we can see that the relationships are multifractal because for different q, there are different exponents; that is, for different q, there are different power–law cross-correlations. Figures 2.6a and 2.6b also shows that for $q = 2$, the cross-correlation scaling exponent $\lambda(q)$ for both the sets CE and DE is greater than 0.5 which means that long-range cross-correlation and persistent properties exist in all the sets (Ghosh et al. 2014) (Figs. 2.6a and 2.6b).

Another evidence of multifractality is shown by the non-linear dependence of the classical multifractal scaling exponent $\tau(q)$ on q which is depicted in Figs. 2.7a and 2.7b, respectively. Further from Fig. 2.8a we can see that the multifractal width of the cross-correlation signal CE is smaller than the multifractal width of the individual set C and set E. Similarly from Fig. 2.8b also we can see that the multifractal width of the cross-correlation signal of set DE is smaller than the multifractal width of the individual set D and set E (Ghosh et al. 2014).

We have mentioned before in Sect. 1.5.1 that the lower value of γ is an indication of higher degree of auto-correlation. Negative value of the auto-correlation γ for set C the data of which are recorded from the hippocampal formation of the opposite hemisphere of the brain in seizure-free interval can be observed from Table 2.2. Drożdż et al. (2009) have also observed negative values of auto-correlation. We further see from Table 2.2 that the degree of auto-correlation γ is least for patients with seizure activity, i.e., for set E (Ghosh et al. 2014).

Table 2.3 shows the degree of cross-correlation γ_x to be the least in set DE where the data of D are recorded from the epileptogenic zone and E from seizure activity of patients. We further observe that the degree of cross-correlation γ_x shows a similar pattern of variation as the degree of multifractality or the scaling exponent $\lambda(q)$,

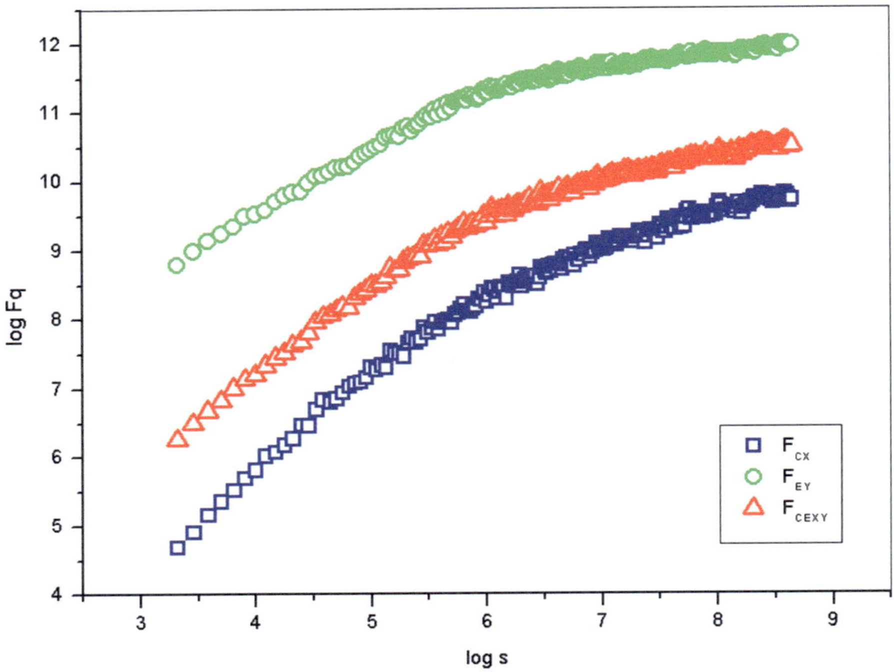

Fig. 2.5a Plot of log F_q vs. log s ($q = 2$) of individual and cross-correlated signal for a particular set of C and E (Ghosh et al. 2014)

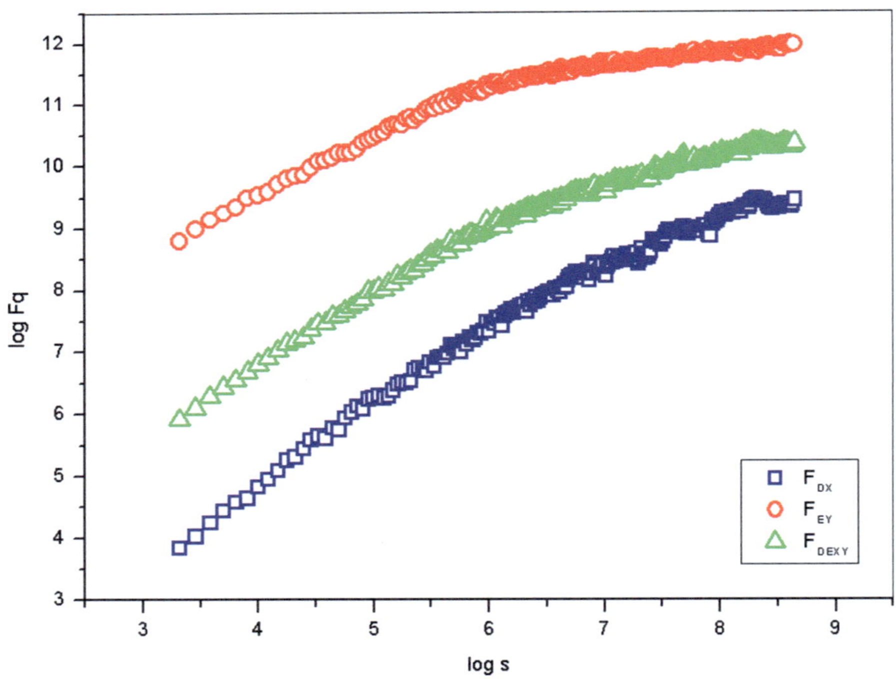

Fig. 2.5b Plot of log F_q vs. log s ($q = 2$) of individual and cross-correlated signal for a particular set of D and E (Ghosh et al. 2014)

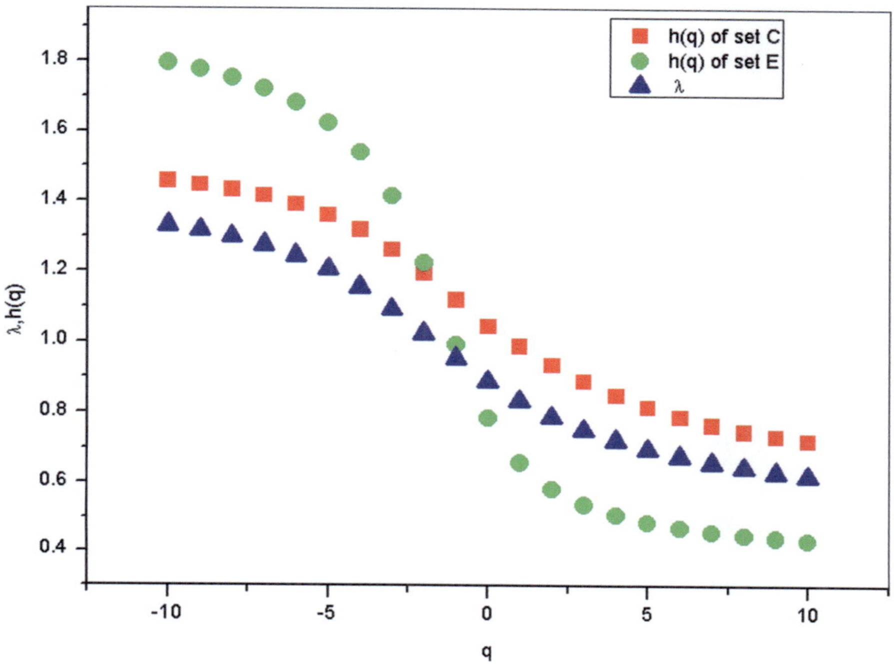

Fig. 2.6a Plot of $\lambda(q)$, $h(q)$ vs. q for a particular signal of set C and E (Ghosh et al. 2014)

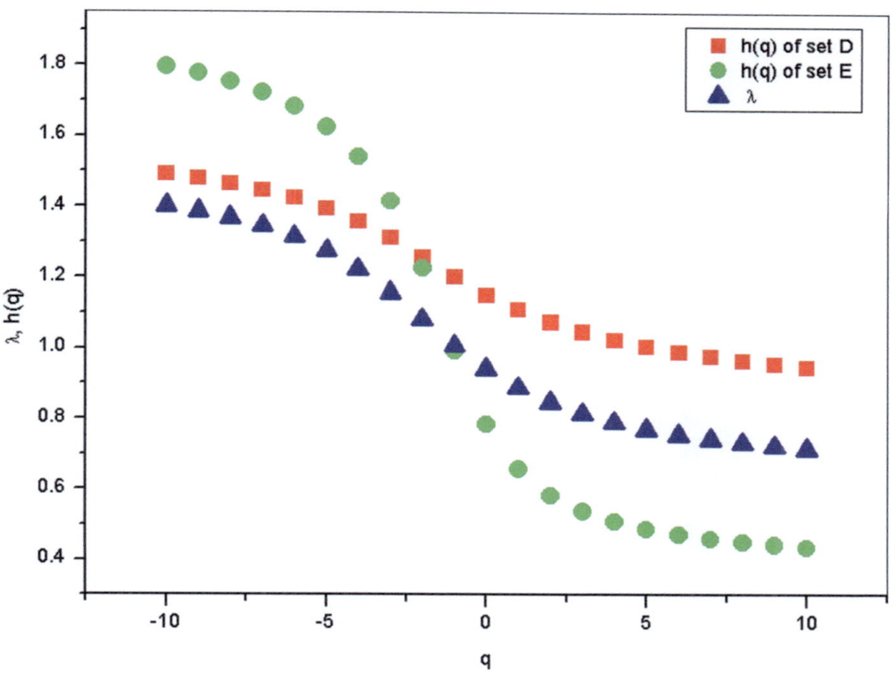

Fig. 2.6b Plot of $\lambda(q)$, $h(q)$ vs. q for a particular signal of set D and E (Ghosh et al. 2014)

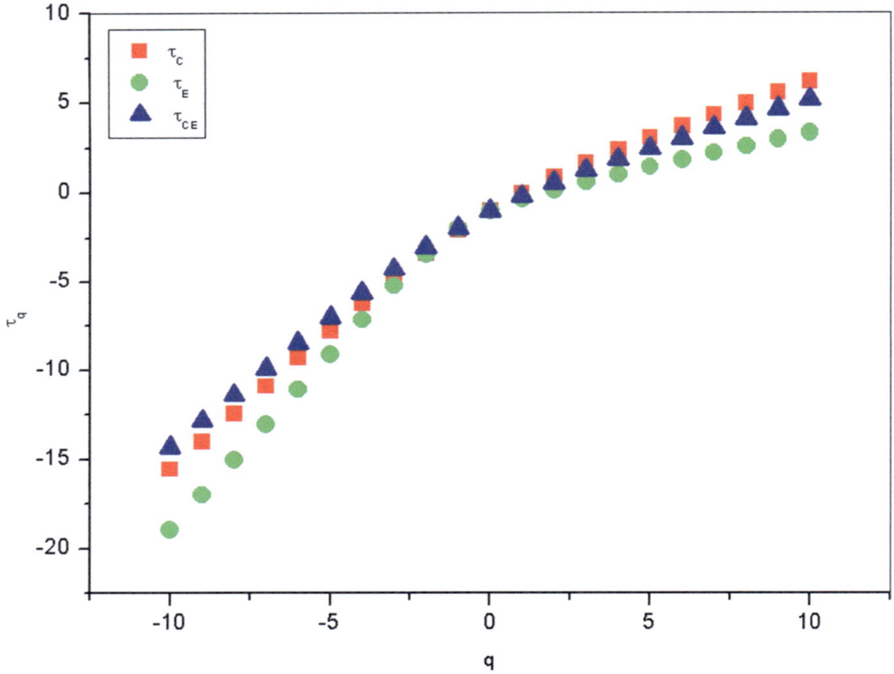

Fig. 2.7a Plot of τ_q vs. q of individual and cross-correlated signal for a particular set of C and E (Ghosh et al. 2014)

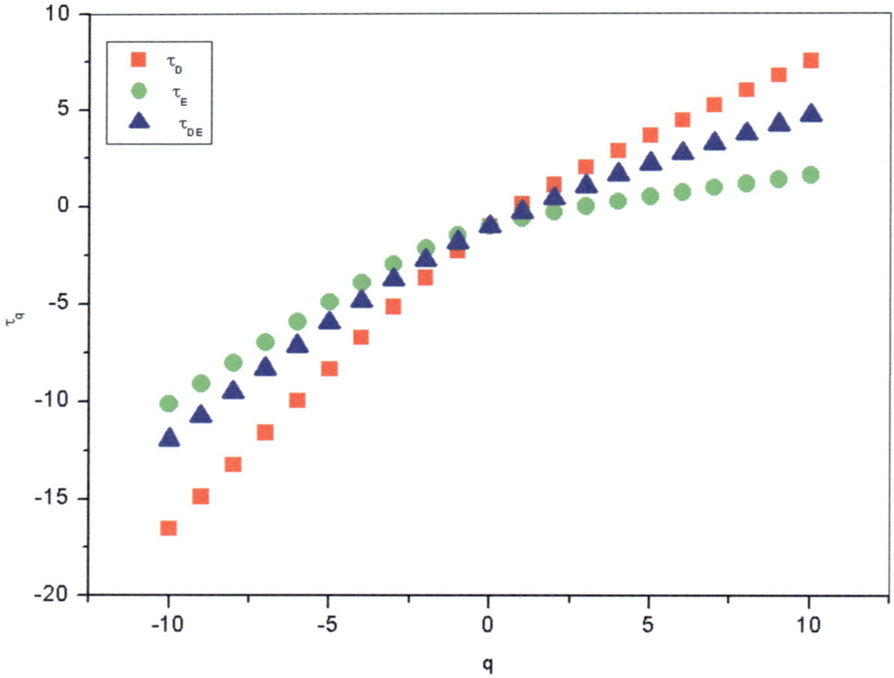

Fig. 2.7b Plot of τ_q vs. q of individual and cross-correlated signal for a particular set of D and E (Ghosh et al. 2014)

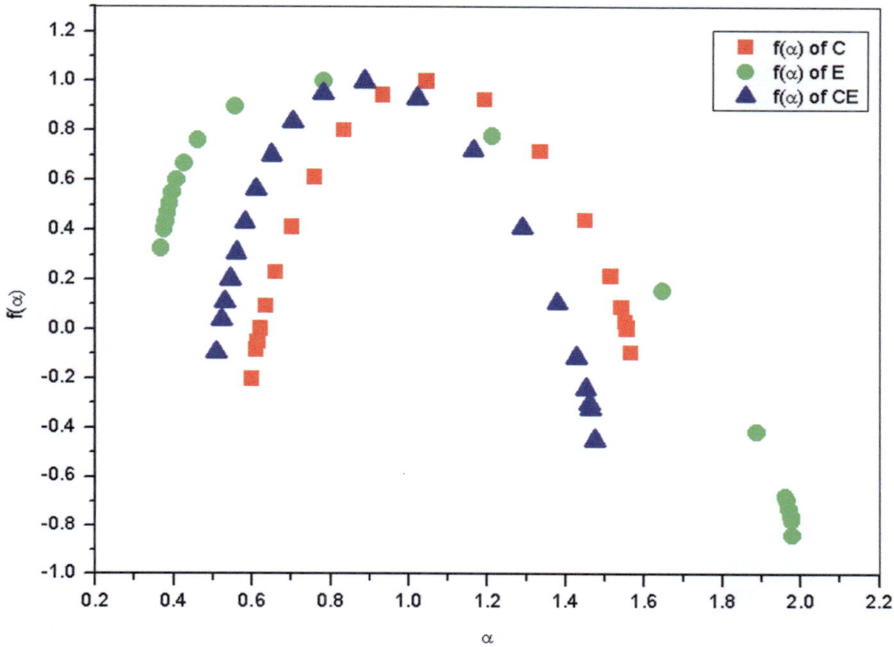

Fig. 2.8a Plot of $f(\alpha)$ vs. α of individual and cross-correlated signal for a particular set of C and E (Ghosh et al. 2014)

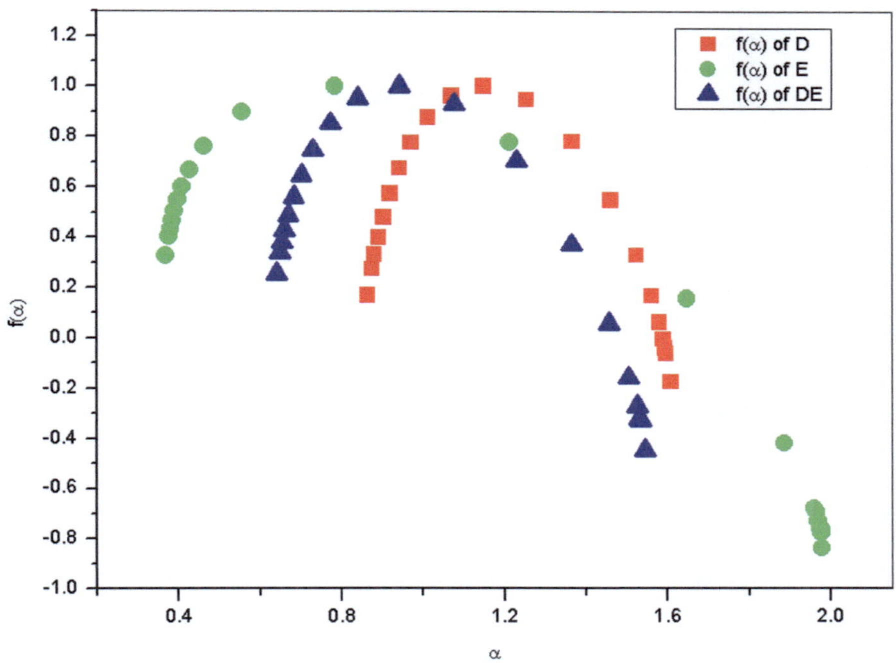

Fig. 2.8b Plot of $f(\alpha)$ vs. α of individual and cross-correlated signal for a particular set of D and E (Ghosh et al. 2014)

Table 2.2 Mean values of h, γ, and W for sets C, D, and E (Ghosh et al. 2014)

Set	Avg $h(q = 2)$	Avg $\gamma(q = 2)$	Avg W
C	1.016	−0.032	0.80
D	0.951	0.095	0.920
E	0.586	0.829	1.190

Table 2.3 Mean values of λ, γ_x, and W for sets CE and DE (Ghosh et al. 2014)

Set	Avg $\lambda(q = 2)$	Avg $\gamma_x(q = 2)$	Avg W
CE	0.811	0.377	0.436
DE	0.789	0.422	0.798

because with decrease in degree of multifractality, the degree of cross-correlation also decreases. Looking at the values of multifractal width (w and w_x) from Tables 2.2 and 2.3, we can observe that the widths of cross-correlation multifractal spectra are narrower than those of separately analyzed sets C, D, and E (Ghosh et al. 2014).

It deserves mentioning that in this investigation the cross-correlation is performed on data sets that are not simultaneous due to the obvious time delay involved which is beyond ones control since this is due to particular features of the data sets. However this cross-correlation study is expected to convey the required information about the correlation (Ghosh et al. 2014).

Thus from the present analysis of EEG signals of epileptic patients using MF-DXA method, we can see that EEG signal during seizure and in seizure-free interval exhibits a power–law cross-correlated behavior. The values of Hurst exponent $h(q = 2)$ and scaling exponent $\lambda(q = 2)$ are an indication of persistent long-range auto- and cross-correlation of the EEG signals. We also observe that among "seizure" and "seizure-free interval," the cross-correlation has larger value in case of epileptogenic zone of the brain. The data is new and important in understanding the dynamics of the disease as well as modeling purpose. The present analysis also suggests to perform similar cross-correlation analysis considering more number of leads as far as practicable which will enrich knowledge for proper diagnosis and prognosis of the disease (Ghosh et al. 2014).

2.5 Possible Application as Biomarker of Epilepsy

As evidenced from this analysis, it is highly effective for diagnosis of the epileptogenic zone. It is needless to mention that this diagnosis is very essential for surgical procedures because the removal of the epileptogenic zone can result in complete cure of the patient. It is also worth mentioning that the multifractal width varies "significantly" for different physiological and pathological cases. Compared to the control group, a complex pattern, assessed with the help of a quantitative parameter, is observed in case of the diseased set. It may be mentioned that in some cases with the conventional technique of assessment of epilepsy with EEG, diagnosis is difficult. On the contrary this method of high precision can not only diagnose epilepsy, but it tells about epileptogenic zone also with simple parameters. Thus a software can easily be developed which can be attached with the EEG system, and eventually this coupled system will be very much effective as so-called biomarker.

Acknowledgment The authors gratefully acknowledge Physica A and Elsevier Publishing Co. for providing the copyrights of Figs. 2.2, 2.3a, 2.3b, 2.3c, and 2.4 and Table 2.1 and Chaos, Solitons, and Fractals for Figs. 2.5a, 2.5b, 2.6a, 2.6b, 2.7a, 2.7b, and 2.8a, 2.8b and Tables 2.2 and 2.3 used in this chapter.

References

Abásolo D, Hornero R, Gómez C, García M, López M (2006) Analysis of EEG background activity in Alzheimer's disease patients with Lempel–Ziv complexity and central tendency measure. Med Eng Phys 28:315–322

Acharya UR, Sree SV, Alvin APC, Yanti R, Suri JS (2012) Application of non-linear and wavelet based features for the automated identification of epileptic EEG signals. Int J Neural Syst 22:1250002

Alberdi A, Aztiria A, Basarab A (2016) On the early diagnosis of Alzheimer's disease from multimodal signals: a survey. Artif Intell Med 71:1–29

Alvarez-Ramirez J, Rodriguez E, Echeverría JC (2005) Detrending fluctuation analysis based on moving average filtering. Phys A 354:199–219

Alzheimer's Association (2017) Alzheimer's disease facts and figures. Alzheimers Dement 13:325–373

Andreu C, de Echave J, Buela-Casal G (1998) Actividad electroencefalográfica según la teoría del caos. Psicothema 10:319–331

Andrzejak RG, Lehnertz K, Mormann F, Rieke C, David P et al (2001) Indications of non-linear deterministic and finite dimensional structures in time series of brain electrical activity: dependence on recording region and brain state. Phys Rev E 64:061907

Austin J, Perkins SM, Johnson C, Fastenau P, Byars A et al (2011) Behaviour problems in children at time of first recognized seizure and changes over the following 3years. Epilepsy Behav 21:373–381

Babiloni C, Pievani M, Vecchio F, Geroldi C, Eusebi F et al (2009) White-matter lesions along the cholinergic tracts are related to cortical sources of EEG rhythms in amnesic mild cognitive impairment. Hum Brain Mapp 30:1431–1443

Baier G, Goodfellow M, Taylor PN, Wang Y, Garry DJ (2012) The importance of modeling epileptic seizure dynamics as spatiotemporal patterns. Front Physiol 3:281

Baker M, Akrofi K, Schiffer R, Boyle MWO (2008) EEG patterns in mild cognitive impairment (MCI) patients. Open Neuroimaging J 2:52–55

Barnes GN, Paolicchi JM (2008) Neuropsychiatric comorbidities in childhood absence epilepsy. Nat Clin Pract Neurol 4:650–651

Bartsch R, Hennig T, Heinen A, Heinrichs S, Maass P (2005) Statistical analysis of fluctuations in the ECG morphology. Phys A 354:415–431

Bashan A, Bartsch R, Kantelhardt JW, Havlin S (2008) Comparison of detrending methods for fluctuation analysis. Phys A 387:5080–5090

Berenyi A, Belluscio M, Mao D, Buzsaki G (2012) Closed-loop control of epilepsy by transcranial electrical stimulation. Science 337:735–737

Bergey GK, Morrell MJ, Mizrahi EM, Goldman A, King-Stephens D et al (2015) Long-term treatment with responsive brain stimulation in adults with refractory partial seizures. Neurology 84:810–817

Blumenfeld H (2012) Impaired consciousness in epilepsy. Lancet Neurol 11:814–826

Bob P, Susta M, Glaslova K, Boutros NN (2010) Dissociative symptoms and interregional EEG cross-correlations in paranoid schizophrenia. Psychiatry Res 177:37–40
Breakspear M (2005) A unifying explanation of primary generalized seizures through non-linear brain modeling and bifurcation analysis. Cereb Cortex 16:1296–1313
Bromfield EB, Cavazos JE, Sirven JI (2006) An introduction to epilepsy. American Epilepsy Society, West Hartford, p 2006
Carron R, Chaillet A, Filipchuk A, Pasillas-Lépine W, Hammond C (2013) Closing the loop of deep brain stimulation. Front Syst Neurosci 7:112
Contreras TI (2007) Análisis Fractal de un sistema complejo: *Epilepsia*. Instituto Politécnico Nacional, Mexico
Curia G, Lucchi C, Vinet J, Gualtieri F, Marinelli C et al (2014) Pathophysiogenesis of mesial temporal lobe epilepsy: is prevention of damage antiepileptogenic? Curr Med Chem 21:663–688
Czigler B, CsikoÂs D, Hidasi Z, Anna Gaal Z, Csibri E et al (2008) Quantitative EEG in early Alzheimer's disease patients – power spectrum and complexity features. Int J Psychophysiol 68:75–80
D'Alessandro M, Vachtsevanos G, Esteller R, Echauz J, Cranstoun S et al (2005) A multi-feature and multi-channel univariate selection process for seizure prediction. Clin Neurophysiol 116:506
Das A, Das P, Roy AB (2002) Applicability of Lyapunov exponent in EEG data analysis. Complex Int 9:1
Dauwels J, Vialatte FB, Weber T, Cichocki A (2009a) Quantifying statistical interdependence by message passing on graphs, Part I: One-dimensional point processes. Neural Comput 21:2152–2202
Dauwels J, Vialatte FB, Weber T, Musha T, Cichocki A (2009b) Quantifying statistical interdependence by message passing on graphs, Part II: Multi-dimensional point processes. Neural Comput 21:2203–2268
Dauwels J, Vialatte FB, Musha T, Cichocki A (2010) A comparative study of synchrony measures for the early diagnosis of Alzheimer's disease based on EEG. NeuroImage 49:668–693
Devinsky O, Vazquez B (1993) Behavioral changes associated with epilepsy. Neurol Clin 11:127–149
Dikanev T, Smirnov D, Wennberg R, Perez Velazquez LJ, Bezruchko BB (2005) EEG nonstationarity during intracranially recorded seizures: statistical and dynamical analysis. Clin Neurophysiol 116:1796
Dojnow P (2007) Comptes rendus de l'Acade'mie bulgare des. Science 60:607
Drożdż S, Kwapien J, Oswiecimka P, Rak R (2009) Quantitative features of multifractal subtleties in time series. Europhys Lett 88:60003
Dutta S (2010a) EEG pattern of normal and epileptic rats: monofractal or multifractal? Fractals 18:425–431
Dutta S (2010b) Multifractal properties of ECG patterns of patients suffering from congestive heart failure. J Stat Mech: Theory Exp:P12021
Dutta S, Ghosh D, Samanta S, Dey S (2014) Multifractal parameters as an indication of different physiological and pathological states of the human brain. Phys A 396:155–163
Easwaramoorthy D, Uthayakumar R (2010) Analysis of EEG signals using advanced generalized fractal dimensions. In: Second international conference on computing, communication and networking technologies, 29–31 July 2010
Escudero J, Sanei S, Jarchi D, AbaÂsolo D, Hornero R (2011) Regional coherence evaluation in mild cognitive impairment and Alzheimer's disease based on adaptively extracted magnetoencephalogram rhythms. Physiol Meas 32:1163–1180
Esteller R, Echauz J, Pless B, Tcheng T, Litt B (2002) Real-time simulation of a seizure detection system suitable for an implantable device. Epilepsia 43(suppl 7):46
Ewers M, Sperling RA, Klunk WE, Weiner MW, Hampel H (2011) Neuroimaging markers for the prediction and early diagnosis of Alzheimer's disease dementia. Trends Neurosci 34:430–442

Falconer K (2003) Fractal geometry: mathematical foundations and applications, 2nd edn. Wiley, Chichester

Fan D, Liu S, Wang Q (2016) Stimulus-induced epileptic spike-wave discharges in thalamocortical model with disinhibition. Sci Rep 6:37703

Fell J, Kaplan A, Darkhovsky B, Roschke J (2000) EEG analysis with non-linear deterministic and stochastic methods: a combined strategy. Acta Neurobiol Exp 60:87–108

Fernández A, Hornero R, Gómez C, Turrero A, Gil-Gregorio P, Matías-Santos J, Ortiz T (2010) Complexity analysis of spontaneous brain activity in Alzheimer disease and mild cognitive impairment: an MEG study. Alzheimer Dis Assoc Disord 24:182–189

Figliola A, Serrano E, Rostas JAP, Hunter M, Rosso OA (2007) Study of EEG brain maturation signals with multifractal detrended fluctuation analysis. AIP Conf Proc 913:190–195

Freeman W, Vitiello G (2006) Non-linear brain dynamics as macroscopic manifestation of underlying many-body field dynamics. Phys Life Rev 3:93–118

Fruend's JE (2003) Chapter 15: Design and analysis of experiments. In: Mathematical statistics with application. Pearson, Boston

Fu K, Qu JF, Chai Y, Zou T (2015) Hilbert marginal spectrum analysis for automatic seizure detection in EEG signals. Biomed Signal Process Control 18:179–185

Gasser US, Rousson V, Hentschel F, Sattel H, Gasser T (2008) Alzheimer disease versus mixed dementias: an EEG perspective. Clin Neurophysiol 119:2255–2259

Gautama T, Mandic DP, Van Hulle M (2003) Indications of non-linear structures in brain electrical activity. Phys Rev E 67:046204

Ghosh D, Dutta S, Chakraborty S (2014) Multifractal detrended cross-correlation analysis for epileptic patient in seizure and seizure free status. Chaos, Solitons Fractals 67:1–10

Gómez C, Hornero R (2010) Entropy and complexity analyses in Alzheimer's disease: an MEG study. Open Biomed Eng J 4:223–235

Gómez C, Hornero R, Abásolo D, Fernández A, Escudero J (2009a) Analysis of MEG background activity in Alzheimer's disease using non-linear methods and ANFIS. Ann Biomed Eng 37:586–594

Gómez C, Mediavilla A, Hornero R, Abásolo D, Fernández A (2009b) Use of the Higuchi's fractal dimension for the analysis of MEG recordings from Alzheimer's disease patients. Med Eng Phys 31:306–313

Gómez C, Martinez-Zarzuela M, Poza J, Diaz-Pernas FJ, Fernandez A, Hornero R (2012) Synchrony analysis of spontaneous MEG activity in Alzheimer's disease patients. In: 2012 annual international conference of the IEEE Engineering in Medicine and Biology Society 2012, pp 6188–6191

Goodfellow M, Schindler K, Baier G (2011) Intermittent spike-wave dynamics in a heterogeneous, spatially extended neural mass model. NeuroImage 55:920–932

Guler NF, Ubeyli ED, Guler I (2005) Recurrent neural networks employing Lyapunov exponents for EEG signals classification. Expert Syst Appl 29:506–514

Guo L, Rivero D, Dorado J, Munteanu CR, Pazos A (2011) Automatic feature extraction using genetic programming: an application to epileptic EEG classification. Expert Syst Appl 38:10425–10436

Gutiérrez J (2001) Detección del foco epiléptico y su ruta de propagación, Memorias II Congreso Latinoamericano de Ingeniería Biomédica. Instituto Nacional de Neurología y Neurocirugía, La Habana

Haghighi HS, Markazi AHD (2017) A new description of epileptic seizures based on dynamic analysis of a thalamocortical model. Sci Rep 7:13615

Harris B, Gath I, Rondouin G, Feuerstein C (1994) On time delay estimation of epileptic EEG. IEEE Trans Biomed Eng 41:820–829

Hassan AR, Siuly S, Zhang Y (2016) Epileptic seizure detection in EEG signals using tunable-q factor wavelet transform and bootstrap aggregating. Comput Methods Prog Biomed 137:247–259

He A, Yang X, Yang Xi, Ning X (2007) Multifractal analysis of epilepsy in electroencephalogram. In: IEEE/ICME international conference on Complex Medical Engineering, 23–27 May 2007

Hornero R, Abasolo D, Escudero J, Gómez C (2009) Non-linear analysis of electroencephalogram and magnetoencephalogram recordings in patients with Alzheimer's disease. Philos Trans R Soc A Math Phys Eng Sci 367:317–336

Houmani N, Vialatte F, Gallego-Jutglà E, Dreyfus G, Nguyen-Michel V, Mariani J, Kinugawa K (2018) Diagnosis of Alzheimer's disease with electroencephalography in a differential framework. PLoS One 13:e0193607

Huang-Jing N, Lu-Ping Z, Peng Z, Xiao-Lin H, Hong-Xing L, Xin-Bao N (2015) Multifractal analysis of white matter structural changes on 3D magnetic resonance imaging between normal aging and early Alzheimer's disease. Chin Phys B 24:070502

Ivanov PC, Amaral LAN, Goldberger AL, Havlin S, Rosenblum MG et al (1999) Multifractality in human heartbeat dynamics. Nature 399:461–465

Ivanov PC, Amaral LAN, Goldberger AL, Havlin S, Rosenblum MG et al (2001) From 1/f noise to multifractal cascades in heartbeat dynamics. Chaos 11:641–652

Ivanov p C, Ma QDY, Bartsch R, Hausdorff JM, Amaral LAN et al (2009) Levels of complexity in scale-invariant neural signals. Phys Rev E 79:041920

Janjarasjit S, Loparo KA (2009) Wavelet-based fractal analysis of the epileptic EEG signal. In: International symposium on intelligent signal processing and communication systems (ISPACS 2009), 7–9 December, pp 127–130

Jelles B, van Birgelen JH, Slaets JPJ, Hekster REM, Jonkman EJ et al (1999) Decrease of non-linear structure in the EEG of Alzheimer patients compared to healthy controls. J Clin Neurophysiol 110:1159–1167

Jeong J, Gore JC, Peterson BS (2001) Mutual information analysis of the EEG in patients with Alzheimer's disease. Clin Neurophysiol 112:827–835

Jeongn J (2002) Non-linear dynamics of EEG in Alzheimer's disease. Drug Dev Res 56:57–66

Jerger KK, Netoff TI, Francis JT, Sauer T, Pecora L, Weinstein SL et al (2001) Early seizure detection. J Clin Neurophysiol 18:259–268

Jing ZL, Lu DZ, Guang HY (2003) Fractal dimension in human cerebellum measured by magnetic resonance imaging. Biophys J 85:4041–4046

Jirsa VK, Stacey WC, Quilichini PP, Ivanov AI, Bernard C (2014) On the nature of seizure dynamics. Brain 137:2210–2230

Jun W, Da-Qing Z (2012) Detrended cross-correlation analysis of electroencephalogram. Chin Phys B 21:028703

Kamath C (2015) Analysis of EEG signals in epileptic patients and control subjects using non-linear deterministic chaotic and fractal quantifiers. Science Postprint 1:e00042

Kannathal N, Acharya R, Alias F, Tiboleng T, Puthusserypady K (2004) Non-linear analysis of EEG signals at different mental states. Biomed Eng Online 3:7

Kannathal N, Rajendra Acharya U, Lim CM, Sadasivan PK (2005) Characterization of EEG—a comparative study. Comput Methods Prog Biomed 80:17–23

Ker MD, Chen WL, Lin CY (2011) Adaptable stimulus driver for epileptic seizure suppression. In IEEE international conference on IC design & technology, 2–4 May 2011

Kim JW, Roberts JA, Robinson PA (2009) Dynamics of epileptic seizures: evolution, spreading, and suppression. J Theor Biol 257:527–532

Kramer MA, Chang FL, Cohen ME, Hudson D, Szeri AJ (2007) Synchronization measures of the scalp EEG can discriminate healthy from Alzheimer's subjects. Int J Neural Syst 17:61–69

Kulish V, Sourin A, Sourina O (2006) Human electroencephalograms seen as fractal time series: mathematical analysis and visualization. Comput Biol Med 36:291–302

Lee SH, Lim JS, Kim JK, Yang J, Lee Y (2014) Classification of normal and epileptic seizure EEG signals using wavelet transform, phase-space reconstruction, and euclidean distance. Comput Methods Prog Biomed 116:10–25

Li Y, Qiu J, Yang Z, Johns EJ, Zhang T (2008) Long-range correlation of renal sympathetic nerve activity in both conscious and anesthetized rats. J Neurosci Methods 172:131–136

Li Y, Wei HL, Billings SA, Liao XF (2012) Time-varying linear and non-linear parametric model for granger causality analysis. Phys Rev E 85:041906

Lin P-J, Neumann PJ (2013) The economics of mild cognitive impairment. Alzheimers Dement 9:58–62

Lin CY, Chen WL, Ker MD (2013) Implantable stimulator for epileptic seizure suppression with loading impedance adaptability. IEEE Trans Biomed Circuits Syst 7:196–203

Litt B, Echauz J (2002) Comparison of three non-linear seizure prediction methods by means of the seizure prediction characteristic. Lancet Neurol 1:22

Lopes da Silva FH, Pijn JP, Boeijinga P (1989) Interdependence of EEG signals: linear vs. non-linear associations and the significance of time delays and phase shifts. Brain Topogr 2:9–18

Lopes da Silva FH, Blanes W, Kalitzin SN, Parra J, Suffczynski P et al (2003a) Dynamical diseases of brain systems: different routes to epileptic seizures. IEEE Trans Biomed Eng 50:540–548

Lopes da Silva FH, Blanes W, Kalitzin SN, Parra J, Suffczynski P et al (2003b) Epilepsies as dynamical diseases of brain systems: basic models of the transition between normal and epileptic activity. Epilepsia 44:72–83

López T, Martínez-González CL, Manjarrez J, Plascencia N, Balankin AS (2009) Fractal analysis of EEG signals in the brain of epileptic rats, with and without biocompatible implanted neuroreservoirs. Appl Mech Mater 15:127–136

Ludescher J, Bogachev MI, Kantelhardt JW, Schumann AY, Bunde A (2011) On spurious and corrupted multifractality: the effects of additive noise, short-term memory and periodic trends. Phys A 390:2480–2490

Lutz A, Greischar LL, Rawlings NB, Ricard M, Davidson RJ (2004) Long-term meditators self-induce high-amplitude gamma synchrony during mental practice. Proc Natl Acad Sci U S A 101:16369

Ma QDY, Bartsch RP, Bernaola-Galvan P, Yoneyama M, Ivanov PC (2010) Effect of extreme data loss on long-range correlated and anticorrelated signals quantified by detrended fluctuation analysis. Phys Rev E 81:031101

Maiwald T, Winterhalder M, Aschenbrenner-Scheibe R, Voss HU, Schulze-Bonhage A, Timmer J (2004) Comparison of three non-linear seizure prediction methods by means of the seizure prediction characteristic. Phys D 194:357

Mann K, Backer P, Roschke J (1993) Dynamical properties of the sleep EEG in different frequency bands. Int J Neurosci 73:161–169

Mars NJ, Lopes da Silva FH (1983) Propagation of seizure activity in kindled dogs. Electroencephalogr Clin Neurophysiol 56:194–209

Marten F, Rodrigues S, Suffczynski P, Richardson MP, Terry JR (2009) Derivation and analysis of an ordinary differential equation mean-field model for studying clinically recorded epilepsy dynamics. Phys Rev E 79:21911

Meghdadi AH, Kinsner W, Fazel-Rezai R (2008) Characterization of healthy and epileptic brain EEG signals by monofractal and multifractal analysis. In: Canadian conference on Electrical and Computer Engineering, June 2008, pp 001407–001411

Milanowski P, Suffczynski P (2016) Seizures start without common signatures of critical transition. Int J Neural Syst 26:1650053

Morales-Matamoros O, Contreras-Troya TI, Mota Hernández CI, Trueba-Ríos B (2009) Fractal analysis of epilepsy. In: Proceedings of the 53rd annual meeting of the international society for the systems sciences, 2009

Mormann F, Kreuz T, Andrzejak RG, David P, Lehnertz K, Elger CE (2003) Epileptic seizures are preceded by a decrease in synchronization. Epilepsy Res 53:173

Mormann F, Andrzejak RG, Elger CE, Lehnertz K (2007) Seizure prediction: the long and winding road. Brain 130:314–333

Murphy JV, Patil A (2003) Stimulation of the nervous system for the management of seizures. CNS Drugs 17:101–115

Nagao M, Murase K, Kikuchi T, Ikeda M, Nebu A et al (2001) Fractal analysis of cerebral blood flow distribution in Alzheimer's disease. J Nucl Med 42:1446–1450

Navarro V, Martinerie J, Quyen MLV, Clemenceau S, Adam C et al (2002) Seizure anticipation in human neocortical partial epilepsy. Brain 125:640

Ni H, Zhou L, Ning X, Wang L (2016) Exploring multifractal-based features for mild Alzheimer's disease classification. Magn Reson Med 76:259–269

Nigam VP, Graupe D (2004) A neural-network-based detection of epilepsy. Neurol Res 26:55–60

Nikulin V, Brismar T (2005) Long-range temporal correlations in electroencephalographic oscillations: relation to topography, frequency band, age and gender. Neuroscience 130:549–558

Ocak H (2009) Automatic detection of epileptic seizures in EEG using discrete wavelet transform and approximate entropy. Expert Syst Appl 36:2027–2036

Osorio I, Frei MG (2007) Hurst parameter estimation for epileptic seizure detection. Commun Inf Syst 7:167–176

Ouyang GX, Li XL, Li Y, Guan XP (2007) Application of wavelet-based similarity analysis to epileptic seizures prediction. Comput Biol Med 37:430–437

Parish L, Worrell GA, Cranstoun SD, Stead SM, Pennell P et al (2004) Long-range temporal correlations in epileptogenic and non-epileptogenic human hippocampus. Neuroscience 125:1069–1076

Park YM, Che HJ, Im CH, Jung HT, Bae SM et al (2008) Decreased EEG synchronization and its correlation with symptom severity in Alzheimer's disease. Neurosci Res 62:112–117

Peiris MTR, Jones RD, Davidson PR, Bones PJ, Myall DJ (2005) Fractal dimension of the EEG for detection of behavioural microsleeps. In: Proceedings of IEEE Engineering in medicine and biology, 27th annual conference Shanghai, China, 1–4 September

Polat K, Güne S (2007) Classification of epileptiform EEG using a hybrid system based on decision tree classifier and fast Fourier transform. Appl Math Comput 187:1017–1026

Poza J, Gómez C, García M, Corralejo R, Fernández A et al (2014) Analysis of neural dynamics in mild cognitive impairment and Alzheimer's disease using wavelet turbulence. J Neural Eng 11:26010

Quyen LVM, Martinerie J, Navarro V, Boon P, Have MD et al (2001) Anticipation of epileptic seizures from standard EEG recordings. Lancet 357:183–188

Rizvi SA, Zenteno JFT, Crawford SL, Wu A (2013) Outpatient ambulatory EEG as an option for epilepsy surgery evaluation instead of inpatient EEG telemetry. Epilepsy Behav Case Rep 1:39–41

Röschke J, Fell J, Beckmann P (1995) Non-linear analysis of sleep EEG in depression: calculation of the largest Lyapunov exponent. Eur Arch Psychiatry Clin Neurosci 245:27–35

Ruiz-Gómez SJ, Gomez C, Poza J, Gutiérrez-Tobal GC, Tola-Arribas MA et al (2018) Automated multiclass classification of spontaneous EEG activity in Alzheimer's disease and mild cognitive impairment. Entropy 20:35

Sackellares JC, Iasemidis LD, Shiau DS, Gilmore RL, Roper SN (2002) Epilepsy—when chaos fails. In: Lehnertz K, Arnhold J, Grassberger P, Elger CE (eds) Chaos in the brain? World Scientific, Singapore, pp 112–133

Salam MT, Perez Velazquez JL, Genov R (2016) Seizure suppression efficacy of closed-loop versus open-loop deep brain stimulation in a rodent model of epilepsy. IEEE Trans Neural Syst Rehabil Eng 24:710–719

Sankari Z, Adeli H, Adeli A (2012) Wavelet coherence model for diagnosis of Alzheimer's disease. Clin EEG Neurosci 43:268–278

Schelter B, Winterhalder M, Maiwald T, Brandt A, Schad A et al (2006) Testing statistical significance of multivariate time series analysis techniques for epileptic seizure prediction. Chaos 16:013108

Serletis D, Bardakjian BL, Valiante TA, Carlen PL (2012) Complexity and multifractality of neuronal noise in mouse and human hippocampal epileptiform dynamics. J Neural Eng 9:056008

Stam CJ (2005) Non-linear dynamical analysis of EEG and MEG: review of an emerging field. Clin Neurophysiol 116:2266–2301

Stam CJ, van Woerkom TCAM, Pritchard WS (1996) EEG measures to characterize EEG changes during mental activity. Electroencephalogr Clin Neurophysiol 99:214–224

Subasi A (2007) EEG signal classification using wavelet feature extraction and a mixture of expert model. Expert Syst Appl 32:1084–1093

Suffczynski P, Kalitzin S, Lopes Da Silva FH (2004) Dynamics of non-convulsive epileptic phenomena modeled by a bistable neuronal network. Neuroscience 126:467–484
Susmakova K (2004) Human sleep and sleep EEG. Meas Sci Rev 4:59–74
Taylor PN, Baier G (2011) A spatially extended model for macroscopic spike-wave discharges. J Comput Neurosci 31:679–684
Taylor PN, Wang Y, Goodfellow M, Dauwels J, Moeller F et al (2014) A computational study of stimulus driven epileptic seizure abatement. PLoS One 9:e114316
Thakor NV, Tong S (2004) Advances in quantitative electroencephalogram analysis methods. Annu Rev Biomed Eng 6:453–495
Timasheva Serge F, Panischev Oleg Y, Polyakov Yuriy S, Demin Sergey A, Kaplan Alexander Y (2012) Analysis of cross-correlations in electroencephalogram signals as an approach to proactive diagnosis of schizophrenia. Phys A 391:1179–1194
Torres NV (1991) Caos en Sistemas Biológicos. Biochemistry and Molecular Biology Department, Santa Cruz de Tenerife
Tzallas AT, Tsipouras MG, Fotiadis DI (2007) Automatic seizure detection based on time-frequency analysis and artificial neural networks. Comput Intell Neurosci 2007:80510
Tzallas AT, Tsipouras MG, Fotiadis DI (2009) Epileptic seizure detection in EEGs using time-frequency analysis. IEEE Trans Inf Technol Biomed 13:703–710
Uthayakumar R, Easwaramoorthy D (2013) Epileptic seizure detection in EEG signals using multifractal analysis and wavelet transform. Fractals 21:1350011
Vingerhoets G (2006) Cognitive effects of seizures. Seizure 15:221–226
Wang J, Niebur E, Hu J, Li X (2016) Suppressing epileptic activity in a neural mass model using a closed-loop proportional-integral controller. Sci Rep 6:27344
Watters PA (2000) Time-invariant EEG power laws. Int J Syst Sci 31:819–826
Watters PA, Martin F (2004) A method for estimating long-range power law correlations from the electroencephalogram. Biol Psychol 66:79–89
Weiss B, Hegedus B, Vago Z, Roska T (2008a) Fractal spectra of intracranial electroencephalograms in different types of epilepsy. In: 19th international EURASIP conference Biosignal, pp 1–5
Weiss B, Vago Z, Tetzlaff R, Roska T (2008b). Long-range dependence of longterm continuous intracranial electroencephalograms for detection and prediction of epileptic seizures. In: international symposium on non-linear theory and its applications, pp 704–707
Wendling F, Hernandez A, Bellanger J, Chauvel P, Bartolomei F (2005) Interictal to ictal transition in human temporal lobe epilepsy: insights from a computational model of intracerebral EEG. J Clin Neurophysiol 22:343–356
Winterhalder M, Maiwald T, Voss HU, Aschenbrenner-Scheibe R, Timmer J et al (2003) The seizure prediction characteristic: a general framework to assess and compare seizure prediction methods. Epilepsy Behav 4:318–325
Winterhalder M, Schelter B, Maiwald T, Brandt A, Schad A et al (2006) Spatio-temporal patient-individual assessment of synchronization changes for epileptic seizure prediction. Clin Neurophysiol 117:2399–2413
Woon WL, Cichocki A, Vialatte F, Musha T (2007) Techniques for early detection of Alzheimer's disease using spontaneous EEG recordings. Physiol Meas 28:335–347
Xu Y, Ma QDY, Schmitt DT, Galvan P, Ivanov PC (2011) Effects of coarse-graining on the scaling behavior of long-range correlated and anti-correlated signals. Phys A 390:4057–4072
Zhang Y, Zhou W, Yuan S (2015) Multifractal analysis and relevance vector machine-based automatic seizure detection in intracranial EEG. Int J Neural Syst 25:1550020
Zhao J, Dou W, Ji H, Wang J (2013) Detrended cross-correlation analysis of epilepsy electroencephalogram signals. In: Proceedings of the 2nd international conference on systems engineering and modeling (ICSEM-13), 2013
Zhou WX (2008) Multifractal detrended cross-correlation analysis for two nonstationary signals. Phys Rev E 77:066211

Chapter 3
Multifractal Approach for Quantification of Autonomic Deregulation Due to Epileptic Seizure with ECG Data

Abstract Since the autonomic function affecting the sympathetic, parasympathetic nervous system changes due to epileptic seizure, the corresponding changes in cardiac signals may be effectively used as potential biomarker providing an extracerebral indicator of ictal onset in some cases. Patients suffering from epilepsy may experience serious cardiac malfunctions leading to sudden unexpected death (SUDEP). Thus the heart rate fluctuations during seizure are non-linear and extremely complex. This chapter presents a report on the analysis of ECG signals of postictal patients using a modern and rigorous chaos-based non-linear technique.

3.1 Introduction

During epilepsy a change in functioning of the cardiac system implies stimulation in the central autonomic network. When a subject encounters seizure, epileptic discharges transmit to the central autonomic network leading to change in normal autonomic control of important cardiac activities. This activation of central autonomic nervous system is the reason behind the peri-ictal autonomic cardiac syndrome seen in epileptic patients. The significance of these autonomic features and the associated complications has been understood clearly in the last few years (Baumgartner et al. 2001; Fogarasi et al. 2006; Widdess-Walsh et al. 2007; Janszky et al. 2007). Thus changes in the pattern of cardiac dynamics in the pre-ictal and ictal stage can act as prospective biomarkers which can be used to develop algorithms to predict seizures (Jensen and Lagae 2010).

Most of the epileptic patients who have suffered sudden unexpected death in epilepsy (SUDEP) have encountered a seizure prior to death (Langan et al. 2000). This has led to the surge of a relationship between seizure and death. An increase in heart rate has been noticed in adults and children with complex partial and generalized tonic–clonic seizures (Marshall et al. 1983; Blumhardt et al. 1986; Smith et al. 1989; Mayer et al. 2004). Blumhardt et al. (1986) reported dominant increase in heart rate of patients with temporal lope seizures. Smith et al. (1989) reported that during the inception of seizure, an initial steep rise is noted in the heart rate of the patients accompanied by distinct variations during the seizure and postictally in a

D. Ghosh et al., *Multifractals and Chronic Diseases of the Central Nervous System*,
https://doi.org/10.1007/978-981-13-3552-5_3

group of patients with complex partial seizures. They also observed similar changes in pattern of heart rate during and after seizure in the same patient which indicates that similar type of autonomic stimulation happened as a standard development in those individuals. In another study Keilson et al. (1989) observed ictal tachycardia of greater than 100 beats per minute in lateralized and generalized seizure patients. Ictal tachycardia in adults patients with refractory temporal lobe seizures have been reported in other studies also (Galimberti et al. 1996; Schernthaner et al. 1999; Garcia et al. 2001; Di Gennaro et al. 2004; Moseley et al. 2011). In comparison to tachycardias (heart rate greater than 100 beats per minute), seizure-related slow asystole and bradycardia (heart rate slower than 60 beats per minute) are less common. An acute slowing of the heart rate giving rise to asystole and syncope is known as ictal bradycardia (Britton et al. 2006). Nashef et al. (1996) reported that ictal bradycardia occurred during respiratory changes, notably apnea, which implies that cardiorespiratory reflexes play a vital role in initiating ictal bradycardia, whereas Tinuper et al. (2001) reported that ictal bradycardia is not dependent on changes due to respiration. Rocamora et al. (2003) reported ictal asystole in very few patients who had undergone video-EEG monitoring. Studies of interictal state of epilepsy were reported by many (Nei 2009).

3.2 Systematic Studies on Abnormalities in Cardiac Autonomic Status

Heart rate and blood pressure assessment while deep breathing and Valsalva activity demonstrates that the operations of the parasympathetic and sympathetic nervous systems are diminished among patients with epilepsy, as compared to control subjects (Isojarvi et al. 1998). Ansakorpi et al. (2000) reported prominent malfunction of cardiovascular autonomic regulation in patients with refractory temporal lobe epilepsy than those of well-controlled temporal lobe epilepsy patients. Ronkainen et al. (2005) reported loss of heart rate variability and subdued circadian rhythm in temporal lope epilepsy and also in controlled and in refractory patients too. Sathyaprabha et al. (2006) measured heart rate and blood pressure at rest, after Valsalva and during change of posture and found autonomic malfunction in patients suffering from chronic refractory epilepsy when compared to control subjects. Their study reported severe malfunction of the autonomic system for long-lasting epilepsy. With the help of simple neurophysiologic tests like Valsalva, tilt table test, RR interval, deep breathing, and sural nerve conductance, Chroni et al. (2008) were able to show that the autonomous nervous system experienced a chronic effect of epilepsy.

Some studies also identify decreased heart rate variability particularly in refractory epilepsy, which demonstrates that changes in the autonomic function may lead to SUDEP (Massetani et al. 1997; Ansakorpi et al. 2002; Persson et al. 2005). Compared to control epilepsy subjects, Hilz et al. (2002) reported diminished variability in the resting heart rate of subjects with SUDEP (Nei et al. 2007). Some studies have also identified changes in autonomic function due to administration of antiepileptic drugs (Isojarvi et al. 1998; Ansakorpi et al. 2000; Hennessy et al. 2001; Daniellson et al. 2005).

With the advancement of computer technology, simultaneous recording of EEG and ECG signals is now possible. Analysis of EEG signals in epileptic patients has come a long way and several clinical and automated methods have been developed till date. Traditionally electroencephalogram tests or brain scans are used to detect epilepsy, but electrocardiogram tests which measure heart rate can also be used as a potential diagnostic tool to detect epilepsy (Su et al. 2008).

Several studies have reported that during seizure along with changes in heart rate, the configuration of the ECG also changes (Varon et al. 2013). For partial and generalized seizures in pre-ictal phase, Zijlmans et al. (2002) reported anomaly in ECG, like T wave inversion and ST elevation/depression. Leutmezer et al. (2003) developed a new method for analyzing peri-ictal heart rate changes which enabled them to access the dynamic changes in ECG during the transition period from interictal to ictal state. Elmpt et al. (2006) modeled heart rate signal using curve-fitting methodology to detect seizure onset from ECG signals. Wong et al. (2008) investigated ECG signals in a seizure clinic and found a close cooperation between cardiology and neurology. Amarnath (2010) used power spectral density (PSD) technique to analyze ECG signals of postictal heart rate oscillations in partial epilepsy. Surges (2010) showed shortening of QT interval during the early postictal phase of refractory temporal lobe epilepsy patients (Ghosh et al. 2017). With a motivation of extracting relevant important information, some other investigations using ECG signals have been reported for both pre-ictal and postictal patients (Naritoku et al. 2003; Sahin et al. 2009; Nilsen et al. 2010; Vanage et al. 2012; Van der Lende et al. 2015). Jansen et al. (2013) reported changes in heart rate in temporal lobe and frontal lobe seizures in childhood epilepsy. Varon et al. (2013) proposed the necessity of developing an easy-to-use alarming system to improve the quality of life of patients suffering from epileptic seizures from the respective changes in heart rate during the pre-ictal, ictal, and postictal phases. The authors used principal component analysis (PCA) to extract features describing the morphology of the ECG signal. They opined that seizure affects the extracted features which can identify when the configuration of ECG mainly that of the QRS-complex is affected. Van der Kruijs et al. (2014) investigated the autonomic nervous system functioning with epileptic seizures in pre-ictal time course of heart rate variability (HRV). Kolsal et al. (2014) have reported a study on heart rate variability in children with epilepsy to predict seizure (Ghosh et al. 2017). In another study Varon et al. (2015) presented two different algorithms, namely, PCA and phase-rectifying signal averaging (PRSA), for determining epileptic seizures from a single-lead ECG. While PCA was shown to capture changes in the morphology of the QRS part in the ECG signal, PRSA quantified the heart rate caused by an epileptic seizure. Qarage et al. (2016) analyzed heart rate variability obtained from ECG employing matching pursuit (MP) and Wigner–Ville distribution (WVD) algorithm to extract important HRV features which can describe seizure and seizure-free status.

ECG being easier to measure and its capacity to note changes in the heart rate prior to onset of disturbances in EEG recordings, ECG is thought to have an in-built superiority over EEG (Varon et al. 2013). But most of the ECG analysis techniques are based on conventional time and frequency domain analysis considering linear fluctuation of heart rate. These techniques are not appropriate to bring about the minute changes in heart rate dynamics (Coenen et al. 1977; Fakhouri 1980; Dasheiff and Dickinson 1986; Oppenheimer 2001; Berilgen et al. 2004; Lahmann et al. 2006; Kamal 2006,2010; De Ferrari et al. 2009; Foldvary-Schaefer and Unnwongse 2011; Meregnani et al. 2011; Zamponi et al. 2011) thus paving the path for introduction of non-linear-based methods to describe complex heart rate dynamics and complement traditional methods of its variability (Kamal 2014). Methods based on non-linear dynamics such as fractal analysis and chaos theory have been introduced (Novak 2011; Previnaire et al. 2012; Moseley et al. 2013). These techniques have been implemented on HR signals and have provided significant clinical information on cardiac diseases (Woo et al. 1994; Brouwer et al. 1996; Huikuri et al. 1996; Mäkikallio et al. 1996,1997; Ho et al. 1997) but have rarely been used on evaluation of autonomic cardiovascular dysfunction in epilepsy (Ghosh et al. 2017).

The pioneer work of Goldberger (1996) introduced the concept of non-linear dynamics in the field of cardiology. Non-linearity of the heart rhythm is an indicator of a patient's overall health. With increasingly sophisticated instruments though, analysis of ECG signals has reached perfection over the years, but the question still remains whether all relevant information contained within the signal can be extracted just by visual inspection. So development of quantitative methods based on non-linear dynamics has gained importance. If the patients' health status can be monitored correctly, then a prediction method can also be developed (Jovic and Bogunovic 2010).

Long-term memory-like structures (Ghosh et al. 2017) are defined by frequency (f) spectrum amplitudes leading to a scale-free power–law relationship of 1/f. Cardiac time series is found to exhibit similar characteristics where the long-range correlations indicate that the fluctuations on one scale are self-similar to those on other scales (Stiedl et al. 2009). Assuming scaling properties of the cardiac time series to be homogeneous throughout the entire signal (Peng et al. 1993,1995; Meyer et al. 1998a, b; Meyer 2002; Meyer et al. 2003; Meyer and Stiedl 2003; Stiedl and Meyer 2003), they were treated as monofractal signals. But later with advancement in analysis techniques, it was revealed that a single scaling parameter is not sufficient to describe the behavior of cardiac time series owing to its inhomogeneous and nonstationary character which is a clear indication that heart rate fluctuations exhibit higher level of multiscale complexity transcending 1/f characteristics (Stiedl et al. 2009). This led to application of multi-exponent multifractal analysis on cardiac time series of healthy controls, patients with cardiac disease, and also study of mice (Ivanov et al. 1999, 2004; Goldberger et al. 2002; Meyer et al. 2003; Meyer and Stiedl 2003, 2006; Stiedl et al. 2004). Thus linear analysis techniques are inappropriate in quantifying dynamics of heart rate given their statistical properties (Stiedl et al. 2009). Non-linear techniques are advanced methods that improve risk cardiovascular stratification in humans (Goldberger 1997; Goldberger et al. 2002; Meyer 2002; Meyer and Stiedl

2003). Non-linear methods being scale-invariant present precise qualitative assessment of heart rate dynamics in both physiological and pathological conditions (Stiedl et al. 2009) with greater sensitivity which linear analyses fail to achieve (Rowan et al. 2007; Montano et al. 2008).

Non-linear and nonstationary character of EEG has been detailed in Chap. 2. Like EEG, ECG signals are also non-linear and nonstationary (Fortrat et al. 1997; Thurner et al. 1998; Pomfrett 1999; Sleight and Donovan 1999; Lin and Hughson 2001; Bernaola et al. 2001; Lass 2002; Yoshikawa and Yasuda 2003). Ivanov et al. (1999) reported healthy human interbeat intervals to exhibit multifractal properties. Amaral et al. (2001) also reported the multifractal behavior of HRV. Wang et al. (2003) too analyzed ECG signals of healthy young adult subjects and old ones and characterized their multifractality (Ghosh et al. 2017). Application of these non-linear techniques in assessing dynamics of the cardiac system motivated us to apply a state-of-the-art methodology to visualize brain function after seizure by studying ECG signals of epileptic patients. An overview of our work is detailed in the following section.

3.3 Multifractal Detrended Fluctuation Analysis of ECG Signals

To provide a complete description of the complicated scaling behaviors of the time series over multiple time scales, Kantelhardt et al. (2002) introduced multifractal detrended fluctuation analysis (MF-DFA) to determine long-range dependence with the crossover time scales. MF-DFA is a widely used methodology for studying different heart diseases (Galaska et al. 2007, 2008; Makowiec et al. 2006, 2007). Galaska et al. (2007) used wavelet transform modulus maxima (WTMM) and MF-DFA to assess correlations between multifractal parameters of heart rate in a group of healthy controls and patients with left ventricular malfunction. In another work Galaska et al. (2008) used the same methodologies (WTMM and MF-DFA) to provide a comparison of the multifractal parameters of heart rate in a group of patients with reduced left ventricular systolic function (rlvs group) and in a group of healthy persons. The width of the multifractal spectrum obtained using MF-DFA was found to be notably reduced, and the Hurst exponent was remarkably higher in rlvs group compared to healthy group both during daytime activity and nighttime rest. Using WTMM width was lower only during diurnal activity. Thus the authors opined MF-DFA to be more precise in contrast to WTMM method for differentiating multifractal properties of the heart rate in control group and patients with left ventricular systolic malfunction. Makowiec et al. (2007) also used WTMM and MF-DFA to study the multifractal properties of power spectra in the low-frequency (LF), very low-frequency (VLF), and ultra-low-frequency (ULF) range. They examined alterations in circadian rhythm for activities performed during the day and while at rest in the night. Normal sinus rhythms during awake and sleep each of 5 h of healthy people were considered. In case of persistence in LF range, loss of heart rate variability at night and also its increase in ULF in the presence of stochasticity are proposed based on qualitative argument.

Wang et al. (2003) used a method developed by Chhabra and Jensen (1989) to compute the multifractal spectrum of ECG signals in young and old subjects. They found multifractal spectrum area in young group to be greater than the old subjects and the logarithm of spectrum area to be inversely proportional with age. The multifractal spectrum area was found to decrease distinctly for brain injury patients. Thus they concluded that multifractal spectrum of ECG is mainly regulated by one's neuro-system. Multifractal spectrum area of men's ECG contemplates the strength of the body's neuroautonomic control on the heart and the extent of heart failure. In old people the impact of neuroautonomic control on the multifractal properties of ECG seemed to be weak, and a decrease in multifractal spectrum was noted. Though both MF-DFA and WTMM have been used in exploring several cardiac disorders, to the best of our knowledge, application of MF-DFA to study ECG signals in epileptic patients post seizure has not been reported. The application of MF-DFA methodology on ECG patterns can help in understanding the changes that occur in heart rate after patients have encountered seizure (Ghosh et al. 2017).

3.3.1 ECG Data

The data for our work which contains seven ECG time series was obtained from "PhysioNet" (https://www.physionet.org/physiobank/database/szdb/) (Al-Aweel et al. 1999). The data consists of 11 partial seizures recorded in 5 women patients, aged between 31 and 48 years, lasting from 15–110 s during continuous EEG, ECG, and video monitoring (Ives et al. 1996). Multiple seizures were recorded for two subjects. The patients had no clinical evidence of cardiac disease and had partial seizures with or without secondary generalization from frontal or temporal foci. The recordings were made under a protocol which was approved by Beth Israel Deaconess Medical Center's (BIDMC) Committee on Clinical Investigations. "Data were analyzed off-line using customized software. Onset and offset of seizures were visually identified to the nearest 0.1 second by an experienced electroencephalographer (DLS) blinded with respect to the HRV analysis. Continuous single-lead ECG signals were sampled at 200 Hz. From the digitized ECG recording, a heartbeat annotation file (a list of the type and time of occurrence of each heartbeat) was obtained using a version of commercially available arrhythmia analysis software" developed by Ho et al. (1997).

The ECG time series of partial seizures recorded in five women patients were analyzed with MF-DFA method. The mathematical details of the method are provided in Appendix A. According to MF-DFA algorithm, each of the time series was first transformed to get the integrated signal which was then divided to N_s bins,

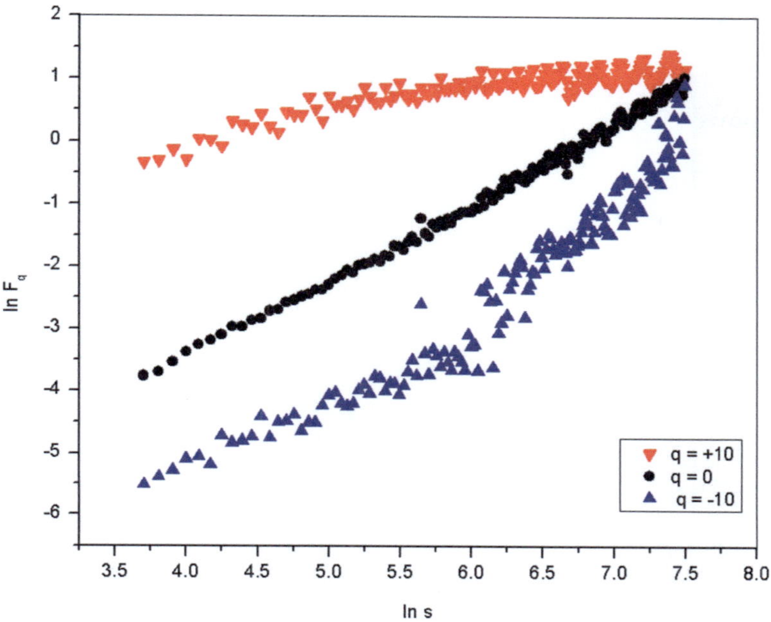

Fig. 3.1 Plot of ln F_q vs. ln s of a particular ECG signal. (Ghosh et al. 2017)

where N_s = int(N/s) (N length of the series, s length of the bin). The fluctuation function $F_q(s)$ was determined for $q = -10$ to $+10$ (q is an index) in steps of 1, using MATLAB code provided by Ihlen (2012). Figure 3.1 depicts plot of the fluctuation of the integrated ECG signals against the length of the bin s (ln $F_q(s)$ versus ln s). Linear dependence of the fluctuation function can be observed which is an indication of scaling behavior of the ECG time series. The values of Hurst exponent $h(q)$ were obtained from the slope of linear fit of ln $F_q(s)$ versus ln s. We know that for a monofractal series, there is only one value of $h(q)$ for all values of q. If $h(q)$ depends on q, the series is multifractal. It has been shown by Kantelhardt et al. (2003) that values of $h(q)$ for $q < 0$ will be more than that for $q > 0$. Thus the values of Hurst exponent $h(q)$ obtained indicate multifractal behavior of the ECG time series. The variation of $h(q)$ for $q < 0$ and $q > 0$ is depicted in Fig. 3.2 which denotes that the degree of multifractality is different in different cases. The values of classical scaling exponent $\tau(q)$ and its variation with q are shown in Fig. 3.3 which is also indicative of multifractal nature of the time series as for a monofractal series, $\tau(q)$ depends linearly on q. Thus the non-linear dependence of $\tau(q)$ on q and the dependence of h (q) on q give evidence for the multifractality of the postictal heart rate oscillations. Figure 3.2 depicts that for $q = 2$, the generalized Hurst exponent $h(q)$ of all the ECG signals is more than 0.5. This indicates that all the signals possess long-range correlation and persistent properties (Ghosh et al. 2017). The multifractal spectrum

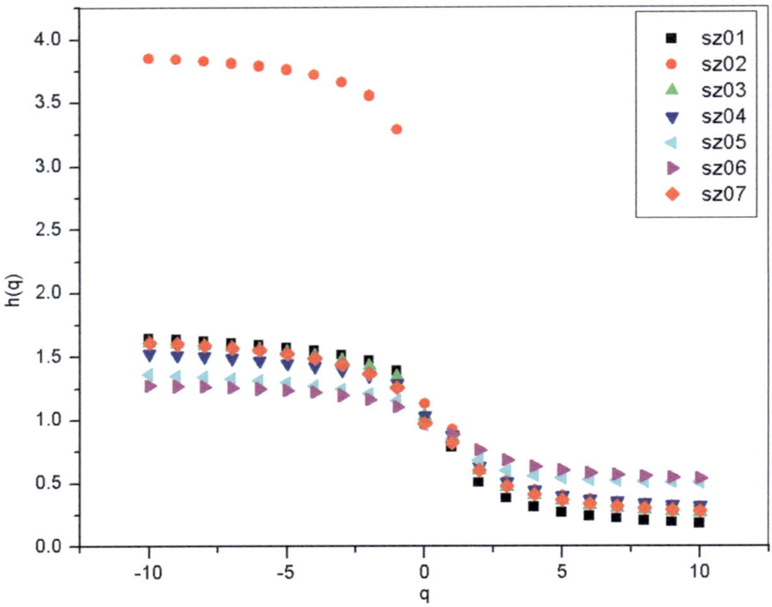

Fig. 3.2 Plot of $h(q)$ vs. q of seven postictal ECG signals. (Ghosh et al. 2017)

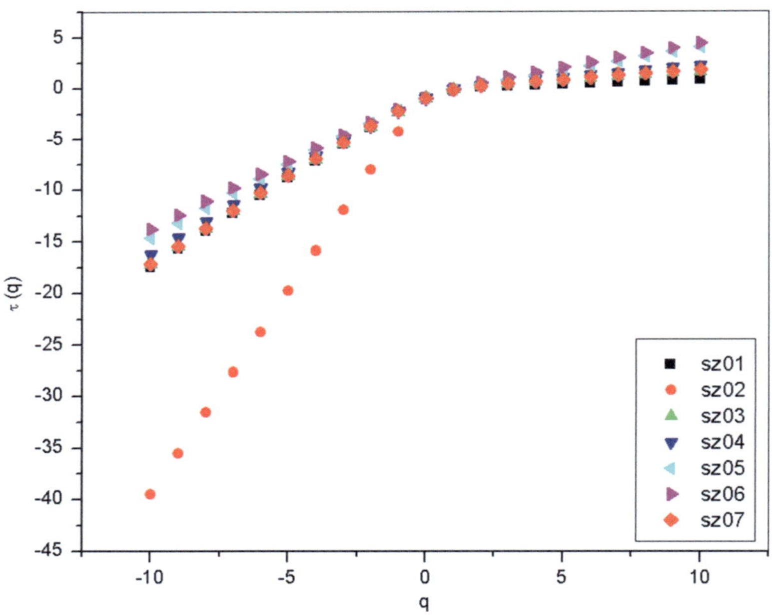

Fig. 3.3 Plot of $\tau(q)$ vs. q of seven postictal ECG signals. (Ghosh et al. 2017)

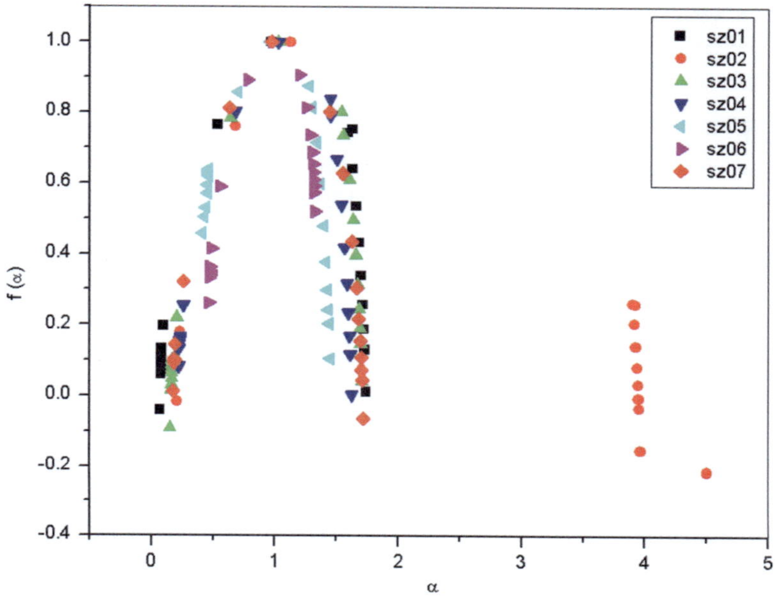

Fig. 3.4 Plot of $f(\alpha)$ vs. α of seven postictal ECG signals. (Ghosh et al. 2017)

Table 3.1 Values of multifractal width (w) and auto-correlation exponent (γ) of seven postictal ECG signals for original and shuffled series (Ghosh et al. 2017)

ECG signals	Multifractal width w		Auto-correlation exponent γ	
	Original	Shuffled	Original	Shuffled
sz01	1.815 ± 0.177	0.894 ± 0.044	0.998 ± 0.012	0.995 ± 0.005
sz02	3.950 ± 0.184	0.498 ± 0.009	0.709 ± 0.012	0.856 ± 0.006
sz03	1.661 ± 0.134	0.781 ± 0.029	0.804 ± 0.014	0.962 ± 0.005
sz04	1.527 ± 0.135	0.654 ± 0.020	0.733 ± 0.012	0.993 ± 0.006
sz05	1.269 ± 0.119	0.761 ± 0.025	0.643 ± 0.007	1.085 ± 0.006
sz06	1.165 ± 0.060	0.403 ± 0.006	0.475 ± 0.007	0.942 ± 0.005
sz07	1.604 ± 0.085	0.742 ± 0.031	0.801 ± 0.006	0.908 ± 0.005

can be used to evaluate the degree of multifractality or complexity in the signal. Ashkenazy et al. (2003) have associated the width of the multifractal spectrum ($f(\alpha)$ versus α) with the degree of multifractality. The multifractal spectrum of seven postictal ECG signals is shown in Fig. 3.4. The values of multifractal width "w" obtained by fitting the multifractal spectrum are shown in Table 3.1. We observe that for all the postictal ECG signals, the multifractal widths are different ranging from as low as 1.17 to as high as 3.95.

Table 3.2 Values of multifractal width (w) of ECG signals of normal healthy people and CHF patients (Channel I) (Ghosh et al. 2017)

ECG signals of healthy people	Multifractal width (w)	ECG signals of CHF patients	Multifractal width (w)
Sample I	1.107 ± 0.152	Sample I	1.735 ± 0.069
Sample II	1.179 ± 0.139	Sample II	2.314 ± 0.087
Sample III	1.090 ± 0.082	Sample III	1.146 ± 0.239
Sample IV	1.073 ± 0.045	Sample IV	2.313 ± 0.039
Sample V	1.110 ± 0.151	Sample V	1.240 ± 0.132

In 2010, Dutta (2010) conducted a study using BIDMC congestive heart failure database with MF-DFA method and reported multifractal width in case of healthy subjects in the range 1.073 to 1.179, whereas for patients suffering from congestive heart failure (CHF), the values correspond to 1.146 to 2.314. The values of multifractal width obtained from BIDMC database are reported in Table 3.2. Thus the values depicted in Tables 3.1 and 3.2 clearly reveal that the multifractal width of ECG recordings of seizure patients is greater than that observed for healthy subjects and in some cases the width of seizure patients ECG exceeds that of CHF also (Ghosh et al. 2017).

From Table 3.1, we can see the value of multifractal width for sz06 is the least and the auto-correlation exponent (γ) which is presented in the same table is 0.48 which reveals a high degree of correlation as we know a lower of value of γ corresponds to a higher degree of correlation. Thus, from these two values of w and γ, we can say that for sz06, the effect of seizure on heart oscillations is the least. Further the same table also reveals the fact that for sz02 the effect of seizure on ECG is the maximum as value of multifractal width is twice than that of rest and γ also approaches uncorrelated behavior which is indicated by value of γ close to 1(Ghosh et al. 2017). To analyze the origin of multifractality, we randomly shuffled the ECG signals and then applied the MF-DFA technique. Difference in values of the multifractal width and auto-correlation exponent for the original and shuffled series are presented in Table 3.1. The shuffled series shows weaker multifractality which implies that the cause of multifractality is due to both long-range correlations and broad probability distribution function. Owing to relatively short sample size, we have not excluded the origin of multifractality due to broad probability distribution function. Values of auto-correlation exponent can be seen as close to 1 in all cases which is an indication that all correlations are spoiled while shuffling the series. For a particular set, the difference in the characteristics of original and shuffled series is depicted in the plots of $h(q)$ vs. q, $\tau(q)$ vs. q, and $f(\alpha)$ vs. α, respectively, in Figs. 3.5, 3.6, and 3.7 (Ghosh et al. 2017).

This study thus clearly states that except sz02, the multifractal width of epileptic patients indicates loss of multifractality which is the outcome of abnormality in the

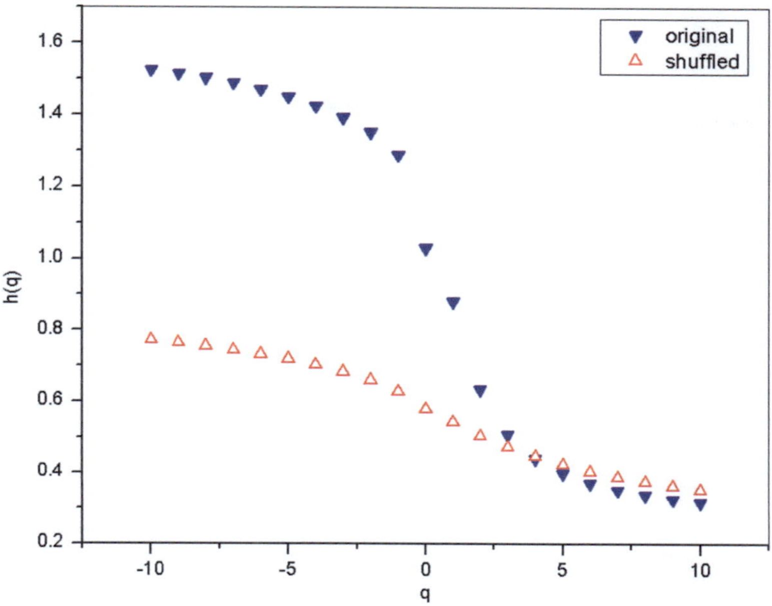

Fig. 3.5 Plot of $h(q)$ vs. q of original and shuffled ECG signal of a particular subject. (Ghosh et al. 2017)

functioning of the heart. Ivanov et al. (1999) also showed loss of multifractality for subjects with a pathological condition-congestive heart failure in contrast to healthy subjects. They observed linearity of classical scaling exponent $\tau(q)$ and narrowness of multifractal spectrum in pathological condition. Same observation was also made by Peng et al. (1995). The case of the patient (sz02) is an uneven one, since contrary to loss of multifractality in other subjects, the present analysis shows an unusual higher degree of multifractality. This observation deserves special attention so far as understanding of dynamics of electrocardiography is concerned. Nevertheless, it can be safely inferred that this anomalous fluctuation has genesis in the epileptic seizure of the patient (Ghosh et al. 2017).

We have reasons to comment that the present analysis of ECG data from postictal patients with a very sensitive and rigorous non-linear technique provides information irrespective of cardiac status of postictal patient quantitatively which is not at all possible with other existing techniques. The present investigation clearly indicates that MF-DFA is a proper tool for further exhaustive investigation taking large data sets which eventually might be able to supply quantitative information about the cardiac status of the patients. This quantitative approach is a step forward toward assessment and monitoring of epileptic patients with the help of quantitative information about the cardiac status.

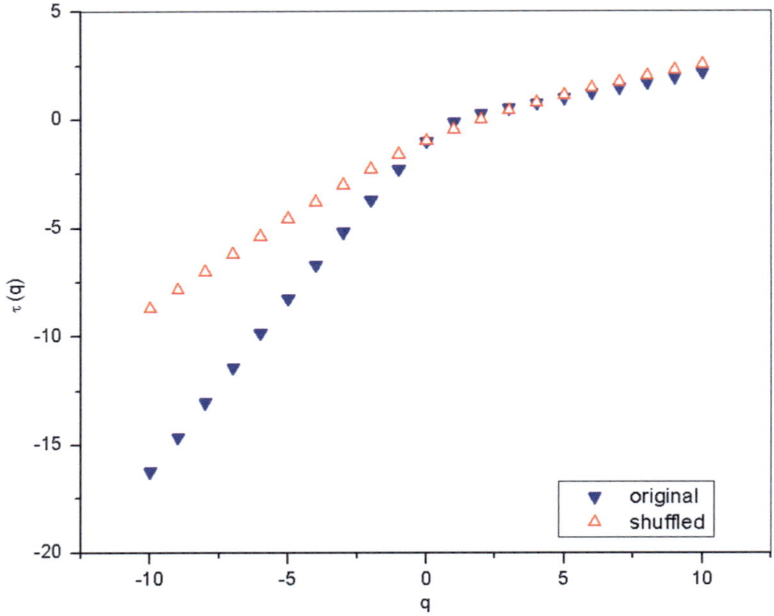

Fig. 3.6 Plot of $\tau(q)$ vs. q of original and shuffled ECG signal of a particular subject. (Ghosh et al. 2017)

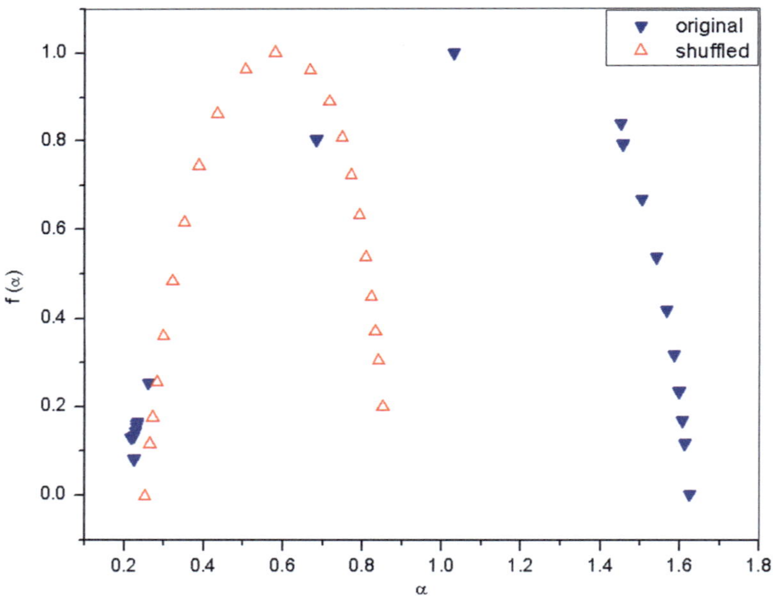

Fig. 3.7 Plot of $f(\alpha)$ vs. α of original and shuffled ECG signal of a particular subject. (Ghosh et al. 2017)

3.4 Results and Possible Biomarker

The application of rigorous non-linear technique in analyzing ECG data of patients clearly supports the fact that the epileptic seizure is associated with the autonomic deregulation. The analysis further shows the degree of autonomic deregulation can be quantified with the help of two parameters, i.e., the multifractal width and autocorrelation exponent.

However, along with postictal data, pre-ictal data for different epileptic patients can be analyzed following this technique which possesses a far-fetching importance for development of software where the findings can be used to develop automatic alarm before seizure as well as even a precursor of cardiac arrest. In this direction since no attempt has been reported so far, the present analysis provides new data using chaos-based latest state-of-the-art methodology which can capture a small change of signal giving rise to a large consequence. It deserves emphasizing that the epilepsy patients experience significant cardiac changes during seizure, causing serious cardiac malfunctions which may lead to SUDEP. Through continuous monitoring of the multifractal parameters, attempts can be made to provide information about the degree of cardiac malfunction for which proper medication can be administered to avoid SUDEP. This method of assessment of ECG signal post seizure, with the help of multifractal parameters as elaborated earlier, can be used to develop a suitable software-based biomarker to help control mortality.

References

Al-Aweel IC, Krishnamurthy KB, Hausdorff JM, Mietus JE, Ives JR et al (1999) Post-ictal heart rate oscillations in partial epilepsy. Neurology 53:1590–1592

Amaral LAN, Ivanov PC, Aoyagi N, Hidaca I, Tomono S et al (2001) Behavioral-independent features of complex heartbeat dynamics. Phys Rev Lett 86:6026

Amaranth M (2010) Diagnosis of post-ictal heart oscillations in partial epilepsy using power spectral density analysis. ICCAE 2nd Int Conf Singapore, IEEE 3:333–336

Ansakorpi H, Korpelainen JT, Suominen K, Tolonen U, Myllyla VV et al (2000) Interictal cardiovascular autonomic responses in patients with temporal lobe epilepsy. Epilepsia 41:42–47

Ansakorpi H, Korpelainen JT, Huikuri HV, Tolonen U, Myllyla VV et al (2002) Heart rate dynamics in refractory and well controlled temporal lobe epilepsy. J Neurol Neurosurg Psychiatry 72:26–30

Ashkenazy Y, Baker DR, Gildor H, Havlin S (2003) Non-linearity and multifractality of climate change in the past 420,000 years. Geophys Res Lett 30:2146–2149

Baumgartner C, Lurger S, Leutmezer F (2001) Autonomic symptoms during epileptic seizures. Epileptic Disord 3:103–116

Berilgen MS, Sari T, Bulut S, Mungen B (2004) Effects of epilepsy on autonomic nervous system and respiratory function tests. Epilepsy Behav 5:513–516

Bernaola GP, Ivanov PC, Amaral LA, Stanley HE (2001) Scale invariance in the non-stationarity of human heart rate. Phys Rev Lett 87:168105–168109

Blumhardt LD, Smith PEM, Owen L (1986) Electrocardiographic accompaniments of temporal lobe epileptic seizures. Lancet 1:1051–1056

Britton JW, Ghearing GR, Benarroch EE, Cascino GD (2006) The ictal bradycardia syndrome: localization and lateralization. Epilepsia 47:737–744

Brouwer J, van Veldhuisen DJ, Veld AJMI (1996) Prognostic value of heart rate variability during long-term follow-up in patients with mild to moderate heart failure. J Am Coll Cardiol 28:1183–1189

Chhabra A, Jensen RV (1989) Direct determination of the f(α) singularity spectrum. Phys Rev Lett *62*:1327–1330

Chroni E, Sirrou V, Trachani E, Sakellaropoulos GC, Polychronopoulos P (2008) Interictal alterations of cardiovagal function in chronic epilepsy. Epilepsy Res 83(2–3):117–123

Coenen AJ, Rompelman O, Kitney RI (1977) Measurement of heart-rate variability: part 2-hardware digital device for the assessment of heart-rate variability. Med Biol Eng Comput 15:423–430

Daniellson BR, Lansdell K, Patmore L, Tomson T (2005) Effects of the antiepileptic drugs lamotrigine, topiramate and gabapentin on hERG potassium currents. Epilepsy Res 63:17–25

Dasheiff RM, Dickinson LJ (1986) Sudden unexpected death of epileptic patient due to cardiac arrhythmia after seizure. Arch Neurol 43:194–196

De Ferrari GM, Sanzo A, Schwartz PJ (2009) Chronic vagal stimulation in patients with congestive heart failure. Conf Proc IEEE Eng Med Biol Soc 2009:2037–2039

Di Gennaro G, Quarato PP, Sebastiano F, Esposito V, Onorati P, Grammaldo LG et al (2004) Ictal heart rate increase precedes EEG discharge in drug-resistant mesial temporal lobe seizures. Clin Neurophysiol 115:1169–1177

Dutta S (2010) Multifractal properties of ECG patterns of patient suffering from congestive heart failure. J Stat Mech Theory Exp P12021

Elmpt WJCV, Nijsen TME, Griep PAM, Arends JBAM (2006) A model of heart rate changes to detect seizures in severe epilepsy. Seizure 15:366–375

Fakhouri SY (1980) Identification of Volterra kernels of non-linear discrete system. IEEE Proc D Contr Theor Appl 127:296–304

Fogarasi A, Janszky J, Tuxhorn I (2006) Autonomic symptoms during childhood partial epileptic seizures. Epilepsia 47:584–588

Foldvary-Schaefer N, Unnwongse K (2011) Localizing and lateralizing features of auras and seizures. Epilepsy Behav 20:160–166

Fortrat JO, Yamamoto Y, Hughson RL (1997) Respiratory influences on non-linear dynamics of heart rate variability in humans. Biol Cybern 77:1–10

Galaska R, Makowiec D, Dudkowska A, Koprowski A, Fijalkowski M et al (2007) Multifractal properties of heart rate by multifractal detrended fluctuation analysis and wavelet transform modulus maxima analysis—are both approaches equivalent? J Electrocardiol 40:S41–S42

Galaska R, Makowiec D, Dudkowska A, Koprowski A, Chlebus K et al (2008) Comparison of wavelet transform modulus maxima and multifractal detrended fluctuation analysis of heart rate in patients with systolic dysfunction of left ventricle. Ann Noninvasive Electrocardiol 13:155–164

Galimberti CA, Marchioni E, Barzizza F, Manni R, Sartori I et al (1996) Partial epileptic seizures of different origin variably affect cardiac rhythm. Epilepsia 37:742–747

Garcia M, D'Giano C, Estelles S, Leiguarda R, Rabinowicz A (2001) Ictal tachycardia: its discriminating potential between temporal and extratemporal seizure foci. Seizure 10:415–419

Ghosh D, Dutta S, Chakraborty S, Samanta S (2017) Epileptic seizure: a new approach for quantification of autonomic deregulation with Chaos based technique. Transl Biomed 8:106

Goldberger AL (1996) Non-linear dynamics for clinicians: chaos theory, fractals, and complexity at the bedside. Lancet 11:1312–1314

Goldberger AL, Amaral LAN, Haussdorff JM, Ivanov PC, Peng CK et al (2002) Fractal dynamics in physiology: alterations with disease and aging. Proc Natl Acad Sci USA 99:2466–2472

Hennessy MJ, Tighe MG, Binnie CD, Nashef L (2001) Sudden withdrawal of carbamazepine increases cardiac sympathetic activity in sleep. Neurology 57:1650–1654

Hilz MJ, Devinsky O, Doyle W, Mauerer A, Dutsch M (2002) Decrease of sympathetic cardiovascular modulation after temporal lobe epilepsy surgery. Brain 125:985–995

Ho KKL, Moody GB, Peng CK, Mietus JE, Larson MG et al (1997) Predicting survival in heart failure cases and controls using fully automated methods for deriving non-linear and conventional indices of heart rate dynamics. Circulation 96:842–848

Huikuri HV, Seppänen T, Koistinen MJ, Airaksinen KEJ, Ikäheimo MJ et al (1996) Abnormalities in beat-to-beat dynamics of heart rate before the spontaneous onset of life-threatening ventricular tachyarrhythmias in patients with prior myocardial infarction. Circulation 93:1836–1844

Ihlen EAF (2012) Introduction to multifractal detrended fluctuation analysis in Matlab. Front Physiol 3:141

Isojarvi JIT, Ansakorpi H, Suominen K, Tolonen U, Repo M, Myllyla VV (1998) Interictal cardiovascular autonomic responses in patients with epilepsy. Epilepsia 39:420–426

Ivanov PC, Amaral LAN, Goldberger AL, Havlin S, Rosenblum MG et al (1999) Multifractality in human heartbeat dynamics. Nature 399:461–465

Ivanov PC, Chen Z, Hu K, Stanley HE (2004) Multiscale aspects of cardiac control. Physica A 344:685–704

Ives JR, Mainwaring NR, Krishnamurthy KB, Blum AS, Drislane FW et al (1996) Technical implementation and clinical findings/results of monitoring oxygen saturation in patients referred for long-term EEG monitoring. Electroencephalogr Clin Neurophysiol 99:432–439

Janszky J, Fogarasi A, Toth V, Magalova V, Gyimesi C et al (2007) Peri-ictal vegetative symptoms in temporal lobe epilepsy. Epilepsy Behav 11:125–129

Jensen K, Lagae L (2010) Cardiac changes in epilepsy. Seizure 19:455–460

Jansen K, Varon C, Huffel SV, Lagae L (2013) Peri-ictal ECG changes in childhood epilepsy: implications for detection systems. Epilepsy Behav 29:72–76

Jovic A, Bogunovic N (2010) Classification of biological signals based on non-linear features. In: Melecon 2010–2010 15th IEEE Mediterranean electrotechnical conference

Kamal A (2006) Assessment of autonomic function for healthy and diabetic patientsUsing entrainment methods and spectral techniques. In: IEEE 32nd annual northeast bioengineering conference, Easton, Pennsylvania, April 1–2, pp 161–162

Kamal AK (2010) Assessment of autonomic function in epileptic patients. Neurosciences (Riyadh) 15:244–248

Kamal AK (2014) Novel method for assessment of autonomic function in health and disease: an application to epilepsy. Int J Neurorehabil 1:133

Kantelhardt JW, Zschiegner SA, Bunde EK, Bunde A, Havlin S et al (2002) Multifractal detrended fluctuation of nonstationary time series. Physica A 316:87–114

Kantelhardt JW, Rybski D, Zschiegner SA, Braun P, Bunde EK et al (2003) Multifractality of river runoff and precipitation: comparison of fluctuation analysis and wavelet methods. Physica A 330:240–245

Keilson MJ, Hauser WA, Magrill JP (1989) Electrocardiographic changes during electrographic seizures. Arch Neurol 46:1169–1170

Kolsal E, Serdaroğlu A, Çilsal E, Kula S, Soysal AS et al (2014) Can heart rate variability in children with epilepsy be used to predict seizures? Seizure 23:357–362

Lahrmann H, Cortelli P, Hilz M, Mathias CJ, Struhal W et al (2006) EFNS guidelines on the diagnosis and management of orthostatic hypotension. Eur J Neurol 13:930–936

Langan Y, Nashef L, Sander JWAS (2000) Sudden unexpected death in epilepsy: a series of witnessed deaths. J Neurol Neurosurg Psychiatry 68:211–213

Lass J (2002) Biosignal interpretation: study of cardiac arrhythmias and electromagnetic field effects on human nervous system (PhD thesis). TTU Press, Tallin

Leutmezer F, Schernthaner C, Lurger S, Pötzelberger K, Baumgartner C (2003) Electrocardiographic changes at the onset of epileptic seizures. Epilepsia 44:348–354

Lin DC, Hughson RL (2001) Modeling heart rate variability in healthy humans: a turbulence analogy. Phys Rev Lett 86:1650–1653
Mäkikallio TH, Seppänen T, Niemelä M, Airaksinen KE, Tulppo M et al (1996) Abnormalities in beat to beat complexity of heart rate dynamics in patients with a previous myocardial infarction. J Am Coll Cardiol 28:1005–1011
Mäkikallio TH, Seppänen T, Airaksinen KEJ, Koistinen J, Tulppo MP et al (1997) Dynamic analysis of heart rate may predict subsequent ventricular tachycardia after myocardial infarction. Am J Cardiol 80:779–783
Makowiec D, Galaska R, Dudkowska A, Rynkiewicz A, Zwierz M (2006) Long-range dependencies in heart rate signals-revisited. Physica A 369:632–644
Makowiec D, Dudkowska A, Galaska R, Rynkiewicz A (2007) Multifractal analysis of normal RR heart-interbeat signals in power spectra ranges. arXiv:q-bio/0702047v1
Marshall DW, Westmoreland BF, Sharbrough FW (1983) Ictal tachycardia during temporal lobe seizures. Mayo Clin Proc 58:443–446
Massetani R, Strata G, Galli R, Gori S, Gneri C et al (1997) Alteration of cardiac function in patients with temporal lobe epilepsy: different roles of EEG/ECG monitoring and spectral analysis of RR variability. Epilepsia 38:363–369
Meregnani J, Clarençon D, Vivier M, Peinnequin A, Mouret C et al (2011) Anti-inflammatory effect of vagus nerve stimulation in a rat model of inflammatory bowel disease. Auton Neurosci 160:82–89
Mayer H, Benninger F, Urak L, Plattner B, Geldner J, Feucht M (2004) EKG abnormalities in children and adolescents with symptomatic temporal lobe epilepsy. Neurology 63:324–328
Meyer M, Marconi C, Ferretti G, Fiocchi R, Skinner JE et al (1998a) Is the heart preadapted to hypoxia? Evidence from fractal dynamics of heartbeat interval fluctuations at high altitude (5050 m). Integr Psychol Behav Sci 33:9–40
Meyer M, Rahmel A, Marconi C, Grassi B, Cerretelli P et al (1998b) Stability of heartbeat interval distributions in chronic high altitude hypoxia. Integr Psychol Behav Sci 33:344–362
Meyer M (2002) Fractal scaling of heart rate dynamics in health and disease. Fractals in biology and medicine, vol 3, Birkhäuser, Basel, pp 181–193
Meyer M, Stiedl O, Kerman B (2003) Discrimination by multifractal spectrum estimation of human heartbeat interval dynamics. Fractals 11:195–204
Meyer M, Stiedl O (2003) Self-affine fractal variability of human heartbeat interval dynamics in health and disease. Eur J Appl Physiol 90:305–316
Montano N, Porta A, Cogliati C, Constantino G, Tobaldini E et al (2008) Heart rate variability explored in the frequency domain: a tool to investigate the link between heart and behavior. Neurosci Biobehav Rev 33:71–80
Moseley BD, Wirrell EC, Nickels K, Johnson JN, Ackerman MJ, Britton J (2011) Electrocardiographic and oximetric changes during partial complex and generalized seizures. Epilepsy Res 95:237–245
Moseley B, Bateman L, Millichap JJ, Wirrell E, Panayiotopoulos CP (2013) Autonomic epileptic seizures, autonomic effects of seizures, and SUDEP. Epilepsy Behav 26:375–385
Naritoku DK, Casebeer DJ, Darbin O (2003) Effects of seizure repetition on postictal and interictal neurocardiac regulation in the rat. Epilepsia 44:912–916
Nashef L, Walker F, Allen P, Sander JW, Shorvon SD, Fish DR (1996) Apnoea and bradycardia during epileptic seizures: relation to sudden death in epilepsy. J Neurol Neurosurg Psychiatry 60:297–300
Nei M, Fertala K, Mintzer S, Skidmore C, Zangaladze A et al (2007) Heart rate and blood pressure in sudden unexpected death in epilepsy. Presented at the annual meeting of the American Academy of Neurology. Boston, MA, May, 2007
Nei M (2009) Cardiac effects of seizures. Epilepsy Curr 9:91–95
Nilsen KB, Haram M, Tangedal S, Sand T, Brodtkorb E (2010) Is elevated pre-ictal heart rate associated with secondary generalization in partial epilepsy? Seizure 19:291–295
Novak P (2011) Quantitative autonomic testing. J Vis Exp 53:2502

Oppenheimer S (2001) Forebrain lateralization and the cardiovascular correlates of epilepsy. Brain 124:2345–2346

Peng CK, Mietus J, Hausdorff JM, Havlin S, Stanley HE et al (1993) Long-range anticorrelations and non-Gaussian behavior of the heartbeat. Phys Rev Lett 70:1343–1346

Peng CK, Havlin S, Stanley HE, Goldberger AL (1995) Quantification of scaling exponents and crossover phenomena in nonstationary heartbeat time series. Chaos 5:82–87

Persson H, Kumlien E, Ericson M, Tomson T (2005) Preoperative heart rate variability in relation to surgery outcome in refractory epilepsy. Neurology 65:1021–1025

Pomfrett CJ (1999) Heart rate variability, BIS and depth of anaesthesia. Br J Anaesth 82:559–661

Previnaire JG, Soler JM, Leclercq V, Denys P (2012) Severity of autonomic dysfunction in patients with complete spinal cord injury. Clin Auton Res 22:9–15

Qarage M, Ismail M, Serpedin E, Zulfi H (2016) Epileptic seizure onset detection based on EEG and ECG data fusion. Epilepsy Behav 58:48–60

Rocamora R, Kurthen M, Lickfett L, Von Oertzen J, Elger CE (2003) Cardiac asystole in epilepsy: clinical and neurophysiologic features. Epilepsia 44:179–185

Ronkainen E, Ansakorpi H, Huikuri HV, Myllyla VV, Isojarvi JI, Korpelainen JT (2005) Suppressed circadian heart rate dynamics in temporal lobe epilepsy. J Neurol Neurosurg Psychiatry 76:1382–1386

Rowan WH 3rd, Campen MJ, Wichers LB, Watkinson WP (2007) Heart rate variability in rodents: uses and caveats in toxicological studies. Cardiovasc Toxicol 7:28–51

Sahin D, Ilbay G, Imal M, Bozdogan O, Ates N (2009) Vagus nerve stimulation suppresses generalized seizure activity and seizuretriggered postictal cardiac rhythm changes in rats. Physiol Res 58:345–350

Sathyaprabha TN, Satishchandra P, Netravathi K, Sinha S, Thennarasu K, Raju TR (2006) Cardiac autonomic dysfunctions in chronic refractory epilepsy. Epilepsy Res 72:49–56

Schernthaner C, Lindinger G, Potzelberger K, Zeiler K, Baumgartner C (1999) Autonomic epilepsy–the influence of epileptic discharges on heart rate and rhythm. Wien Klin Wochenschr 111:392–401

Sleight JW, Donovan J (1999) Comparison of bispectral index, 95% spectral edge frequency and approximate entropy of the EEG, with changes in heart rate variability during induction of general anaesthesia. Br J Anaesth 82:666–671

Smith PEM, Howell SJ, Owen L, Blumhardt LD (1989) Profiles of instant heart rate during partial seizures. Electroencephalogr Clin Neurophysiol 72:207–217

Stiedl O, Jansen RF, Pieneman AW, Ögren SO, Meyer M (2009) Assessing aversive emotional states through the heart in mice: implications for cardiovascular dysregulation in affective disorders. Neurosci Biobehav Rev 33:181–190

Stiedl O, Meyer M (2003) Fractal dynamics in circadian time series of cortico-tropin-releasing factor subtype 2-deficient mice. J Math Biol 47:169–197

Stiedl O, Tovote P, Ögren SO, Meyer M (2004) Behavioral and autonomic dynamics during contextual fear conditioning in mice. Auton Neurosci 115:15–27

Su ZY, Wu T, Yang PH, Wang YT (2008) Dynamic analysis of heartbeat rate signals of epileptics using multidimensional phase space reconstruction approach. Physica A 387:2293–2305

Surges R (2010) Enhanced QT shortening and persistent tachycardia after generalized seizures. Neurology 74:421–426

Thurner S, Feurstein MC, Teich MC (1998) Multiresolution wavelet analysis of heartbeat intervals discriminates healthy patients from those with cardiac pathology. Phys Rev Lett 80:1544–1547

Tinuper P, Bisulli F, Cerullo A, Carcangiu R, Marini C et al (2001) Ictal bradycardia in partial epileptic seizures: autonomic investigation in three cases and literature review. Brain 124:2361–2371

Vanage AM, Khade RH, Shinde DB (2012) Classifying five different arrhythmias by analyzing the ECG signals. IJCEM 15:2230–7893

Van der Kruijs SJM, Vonck KEJ, Feijs LMG, Bodde NMG, Lazeron RHC et al (2014) Autonomic nervous system functioning associated with epileptic seizures: analysis of heart rate variability. J Neurol Neurophysiol 5:215–217

Van der Lende M, Surges R, Sander JW, Thijs RD (2015) Cardiac arrhythmias during or after epileptic seizures. J Neurol Neurosurg Psychiatry 10:1–6

Varon C, Jansen K, Lagae L, Huffel SV (2013) Detection of epileptic seizures by means of morphological changes in the ECG. In: Computing in cardiology conference, pp 8630–8866

Varon C, Jansen K, Lagae L, Huffel SV (2015) Can ECG monitoring identify seizures? J Electrocardiol 48:1069–1074

Wang J, Ning X, Chen Y (2003) Multifractal analysis of electronic cardiogram taken from healthy and unhealthy adult subjects. Physica A 323:561–568

Widdess-Walsh P, Kotagal P, Jeha L, Wu G, Burgess R (2007) Multiple auras: clinical significance and pathophysiology. Neurology 69:755–761

Wong SH, Adams P, Jackson M (2008) The electrocardiograph (ECG) in a first seizure clinic. Seizure 17:707–710

Woo MA, Stevenson WG, Moser DK, Middlekauff HR (1994) Complex heart rate variability and serum norepinephrine levels in patients with advanced heart failure. J Am Coll Cardiol 23:565–569

Yoshikawa Y, Yasuda Y (2003) Non-linear dynamics in heart rate variability in different generations. Bull Toyohashi Sozo Coll 7:63–78

Zamponi N, Passamonti C, Cesaroni E, Trignani R, Rychlicki F (2011) Effectiveness of vagal nerve stimulation (VNS) in patients with drop-attacks and different epileptic syndromes. Seizure 20:468–474

Zijlmans M, Flanagan D, Gotman J (2002) Heart rate changes and ECG abnormalities during epileptic seizures: prevalence and definition of an objective clinical sign. Epilepsia 43:847–854

Chapter 4
Multifractal Analysis of Electromyography Data

Abstract Myopathies (MYO) are a group of disorders where malfunction of muscle fibers occurs for a number of reasons which results in a muscular dysfunction manifesting weakness of muscles. Neuropathies are also disorders of the peripheral nervous system for which information transmission from brain and spinal cord to every other part of the body is disturbed. For diagnosis and characterization of motor neuron disease (MND), myopathy, and neuropathy, the electromyography (EMG) is extensively used since EMG signal can be analyzed to obtain information in regard to degree of disorder. The contents of the chapter deal with the details of a rigorous and robust non-linear technique, namely, multifractal detrended fluctuation analysis, to assess the multifractal property of EMG signals of patients with neuromuscular disorders and also use of two quantitative parameters, the multifractal width, and the auto-correlation exponent as biomarker for diagnosis and prognosis of both MYO and NEURO and even for early detection of MND.

4.1 Introduction

The human skeletal muscular system is comprised of the nervous system and the muscular system, which collectively is known as the neuromuscular system. The disorders which originate in the nervous system, in the neuromuscular junctions, and in the muscle fibers are known as neuromuscular disorders. The disease may cause minor of strength or its severity may lead to even amputation resulting from neuron or muscle death (Subasi 2013). Proper diagnosis of the disorder is crucial so that treatment can begin at an early stage (Alkan and Gunay 2012). Among several neuromuscular disorders, we have considered myopathy, neuropathy, or amyotrophic lateral sclerosis (ALS).

D. Ghosh et al., *Multifractals and Chronic Diseases of the Central Nervous System*,
https://doi.org/10.1007/978-981-13-3552-5_4

4.2 Motor Neuron and Musculoskeletal Disease: Neuropathy and Myopathy

Neuropathy (NEURO) or amyotrophic lateral sclerosis (ALS) is a relatively rare neurodegenerative disorder of the peripheral nervous system for which information transmission from brain and spinal cord to every other part of the body is disturbed. Some typical symptoms are short-term insensibility, tingling and pricking sensations, reactions to touch, and even weakness in the muscles. Muscle wasting, burning pain, paralysis, and organ or gland dysfunctions are other symptoms of neuropathy. Longer duration of motor unit potentials with increased amplitude is associated with neuropathy (Goen 2014). This disease is a part of motor neuron disorder in which loss of life is the ultimate.

Myopathy (MYO) is a pathological condition where primarily the skeletal muscle fibers are affected. Muscle dysfunction, cramps, stiffness, and spasm are some of the common symptoms of myopathy. They can be either inherited or acquired. Most muscular dystrophies are hereditary, causing severe degenerative changes in the muscle fibers. Polymyositis is a frequently encountered myopathy associated with commencement of weakening of the muscles which progresses slowly. In myopathy patients' short-term motor unit potentials with reduced amplitude are noticed (Trojaborg 1987).

For diagnosis of both disorders, patients are first interviewed by doctors, but in some cases they feel so weak that they are not able to even speak (Artameeyanant et al. 2016). Under such circumstances electromyography (EMG) of the muscle signals are recorded which can be diagnosed by a neurological expert to detect myopathy and neuropathy (ALS) (Kincaid 2015; Weiss et al. 2015; Gitiaux et al. 2016).

4.3 Electromyography – A Tool to Detect Motor Neuron Disease

Electromyography (EMG) is a standard technique to determine neurophysiologic characteristics of skeletal muscles to diagnose neuromuscular disorders. EMG is the record of electrical action potentials originated from a group of muscle fibers. These muscle fibers are governed by the same motor nerve and are called motor unit. These motor units which are the basic units of the muscle can be activated voluntarily. The structure and condition of a motor unit can be assumed from the shape of waveform of individual motor unit action potential (MUAP) (Bue et al. 2013). The motor unit action potentials (Nikolic and Krarup 2011) have significant role in diagnosis of disease and are extensively used by neurophysiologists. For abnormal muscles the structure of MUAPs may become different which can be helpful in the identification of disease (Fuglsang-Frederiksen 2000). Since manual inspection of the complicated structures of MUAPs is difficult to assess and time consuming, hence detection of abnormalities in structure of MUAPs can be done quantitatively (Sharma et al. 2017). We have already discussed in Chap. 1 (Sect. 1.2.3) about the types of EMG – the SEMG and IEMG (Farina and Negro 2012).

The movement or change in posture of any part of the body is associated with the strength the strength of SEMG. The working condition of the muscle fibers is reflected in the SEMG signals (Basmajian and De Luca 1985). Intramuscular EMG (Monsifrot et al. 2004) which is an invasive method is not commonly used due to the pain involved in the process. In this method a fine-wire or needle-type electrode is introduced through the skin which reaches deep down the muscles. Advancement of diagnosis technique for evaluation of neuromuscular disorders based on EMG records is important and need of the day (Gokgoz and Subasi 2015).

4.4 Study of SEMG Signals

The surface electromyography (SEMG) signal has broad application in the areas of neuromuscular disorder, rehabilitation, and control of prosthetic devices including man–machine interface of individuals with amputations or congenitally deficient limbs (Hudgins et al. 1993; Kang et al. 1996; Chang et al. 1996; Goge and Chan 2004; Acharya et al. 2011; Subasi 2013; Riillo et al. 2014). Changes in the properties of the EMG signal signify neuromuscular disorder or other pathological conditions originating either in the nervous system or in the muscles (Lima et al. 2016). Different linear analysis methods have been used to describe the characteristics of EMG signals. Time-domain features have been studied by zero crossings and root mean square (RMS) (Hudgins et al. 1993) techniques, stochastic features by autoregressive model coefficients (Farina et al. 2001), cepstral coefficients (Chang et al. 1996), mean frequency and median frequency (MDN) (Gerdle and Eriksson 1990; Kang et al. 1996), etc. But there are certain limitations with these methods (Ghosh et al. 2017).

Like other bioelectrical signals, EMG is also innately non-linear in nature. Some of the typical characteristics it possesses are scale invariance, scaling range, power–law scaling, and self-similarity (Eke et al. 2002; Sarkar and Leong 2003). We know that in self-similarity a small-scale structure is a replica of the large-scale structure of an object. Self-similarity can not only characterize different biomedical signals but can also identify diverse arrangements in these signals (Janjarasjitt 2014; Najarian and Splinter 2012). Self-similarity in EMG signals can be described by fractal dimension (FD) measures (Hu et al. 2005), presenting the means to abstract distinct features directly from these signals (Easwaramoorthy and Uthayakumar 2011). Anmuth et al. (1994) and Gupta et al. (1997) have demonstrated the fractal nature of SEMG. They noticed that as value of FD changes, the fraction of maximum voluntary contraction (MVC) and flexion–extension speeds also changes. Webber et al. (1995) used recurrence quantification analysis (RQA) to investigate the changes of the percent determinism (%DET) in the course of muscle fatigue and observed that the %DET measure performed significantly better than the (median frequency) MDF. A relation was found to exist between FD of needle EMG signal and the strength of muscle contraction by Gitter and Czerniecki (1995). Several other studies too demonstrated the FD of SEMG signals (Xu and Xiao 1997; Chen and Wang 2000; Hu et al. 2005). In a study conducted by Nussbaum and Yassierli (2003) among other non-linear methods used for characterizing EMG signals, fractal dimension was found to be sensitive to the magnitude and rate of the generated muscle force. To determine minute changes in the force of contraction, Ravier et al. (2005) identified that the right slope index (RSI) of the fractal curve of SEMG has a relation with the force of contraction. Arjunan and Kumar (2007) used the fractal theory to estimate the level and source of muscle activity of different hand gestures for determining the corresponding hand gesture. They found FD to define the complexity of the SEMG signal and maximum fractal length (MFL) represented the extent of activity in the muscles. When multiple muscles are functioning, this feature is useful for analysis and modeling of SEMG patterns. Thus based on these studies, Arjunan and Kumar (2014) inferred that fractal analysis of SEMG can be used to determine the force of contraction of the muscle. To describe the irregularity of a statistically stationary signal, fractal dimension is an important tool as it defines the scale-invariant non-linear property of the signal. Arjunan and Kumar (2014) opined that since small change in the muscle activity does not affect FD and signal spectra of SEMG signals, thus FD should be the measure of muscle property such as size and not muscle activity (Fox 1989; Kupa et al. 1995; Gabriel and Kamen 2009). From this Arjunan and Kumar (2014) hypothesized that FD is not sufficient to describe the force of contraction of muscles. The change in the value of FD due to the increase in force of muscle contraction as reported in some works (Anmuth et al. 1994; Gupta et al. 1997; Xu and Xiao 1997; Hu et al. 2005) is the result of muscle size change that is related to high levels of muscle contraction. Arjunan and Kumar (2014) identified a new fractal-based feature (maximum fractal length – MFL) of SEMG that relates to the strength of the muscle contraction even when contraction is just 20%. Compared to other methods like as variance (VAR), waveform length (WL), RSI, FD, and root mean square (RMS), the authors argued that both the experimental results and linear regression analysis demonstrate MFL to be a better estimate of strength of muscle contraction which can even detect minute muscle activity changes. Ancillao et al. (2014) conducted an experiment to investigate the correlation among the fractal dimension of the surface EMG signal obtained the main erector muscle of the human leg, during a vertical jump and the height reached in that jump. FD was found to characterize the EMG signal and significantly high correlation coefficient between fractal dimension and height of the jump was obtained by linear regression analysis in all the healthy subjects.

Several other works using different non-linear methods to determine the geometry and fractal features of EMG signals (Chang et al. 2000, 2004, 2007; Chang and Chang 2002; Chen et al. 2006; Shields 2006) have been reported. To evaluate the fractal dimension, geometrical methods like Katz method (1988) and box-counting method (Falconer 2003) have been applied by some authors. Other non-linear methods such as non-linear entropy analysis have been applied on SEMG signals to gather information about changes in different muscle statuses (Ghosh et al. 2017).

For analyzing the fingers movements from three channels of SEMG Zhao et al. (2007) employed sample entropy and wavelet transform coefficients. Naik et al. (2009) made use of fractal dimension features to identify finger movements. Gang et al. (2007) observed multifractality of SEMG signals obtained from biceps brachii on the skin surface of right forearm of human subjects during a static contraction. With DFA Phinyomark et al. (2009) determined the self-similarity of SEMG signal and derived features different from mean absolute value (MAV) or root mean square (RMS). In another work Phinyomark et al. (2010) once again demonstrated the usefulness of DFA method for characterizing SEMG signals. The fractal scaling exponent obtained can be a helpful tool that can characterize SEMG signals of prosthetic or robot arm. In 2012 Phinyomark et al. (2012) identified DFA to be a better measure compared to conventional techniques in characterizing EMG signals from bifunctional movements, like flexion–extension. Principal component analysis (PCA) allowed the examining of EMG activation relation among a large number of muscles during lifting task (Moreside et al. 2014). Dang et al. (2012) showed EMG to be a powerful tool for investigating the relationship between jaw imbalance and the loss of arm strength with Higuchi's fractal dimension (HFD) analysis. Lei and Meng (2012) investigated the stochastic, deterministic, and chaotic behavior of SEMG signals with non-linear techniques like surrogate data method, Volterra–Wiener–Korenberg (VWK) model method, chaotic analysis method, symplectic geometry method, and fractal analysis method. Difficulty in describing SEMG using single fractal dimension, the authors understood the necessity of multifractal analysis. For categorizing myopathy and healthy subjects using EMG signal, Patidar et al. (2013) applied the back-propagation neural network classifier. Naeem et al. (2014) used linear and non-linear techniques in combination to estimate their ability to identify uterine EMG signals of term and preterm deliveries using artificial neural network.

Vishnu and Shalu (2015) developed a technique to equate SEMG signals recorded from biceps and triceps muscles with the corresponding angular velocity of motion of forearm with a linear model – the auto-regressive exogenous input (ARX) and two non-linear models, namely, Hammerstein and artificial neural network (ANN). The results showed that "SEMG–angular velocity" model based on ANN performs better in contrast to the conventional parametric system identification models like Hammerstein and ARX. Several other authors have used ANN to quantify EMG signals. Earlier Koike and Kawato (1995) used EMG signals recorded from single-joint isometric motions to decode arm motion using a neural network-based model. Smith et al. (2008) used ANN to study the continuous motion of the fingers, using myoelectric activity from muscles of the forearm. Ryu et al. (2008) also used neural network to study EMG signals of a one degree of freedom (DoF) robot arm. Some other studies using the state–space model have been used to evaluate human arm kinematics from myoelectric activity produced by certain muscle groups (Artemiadis and Kyriakopoulos 2007, 2010, 2011). To characterize EMG signals, Mishra et al. (2016) and Naik et al. (2016) used empirical mode decomposition (EMD) which can decode signals from data collected in noisy non-linear and nonstationary processes (Artameeyanant et al. 2016). Lima et al. (2016) used relevance vector machines (RVM) and fractal dimension (FD) for automatically identifying EMG signals related to different classes of limb motion. Built on normalized weight vertical visibility algorithm (NWVVA), an EMG feature extraction technique was presented by Artameeyanant et al. (2016) to detect healthy, ALS, and myopathy statuses.

Fele et al. (2008) made a comparison among different linear and non-linear techniques to differentiate EMG records of term and preterm delivery groups. Non-linear techniques were found to be more suitable in discriminating the two groups. Some works have argued that for clinical purposes, non-linear correlation coefficient can enhance the utility of uterine EMG signals (Jezewski et al. 2005; Khalil and Duchene 2007; Hassan et al. 2007, 2012; Lucovnik et al. 2011; Alamedine et al. 2013). For determining uterine EMG signals during pregnancy and labor, Diab et al. (2010, 2012) also applied non-linear analysis techniques. Ren et al. (2015) were of the opinion that from data of uterine EMG can help in predicting risk of preterm delivery. To analyze uterine EMG signals, various signal processing techniques have been applied. They opined that using EMD method preterm and term delivery records based on the entropy ratios of the instantaneous amplitude and frequency of each two IMFs of uterine EMG signals can be classified. Thus analysis of uterine EMG signals is important to help prevent premature birth thus decreasing risk of infant mortality.

With a view to characterize neuromuscular disorders, we studied EMG signals obtained from one healthy and diseased subjects using a non-linear method the multifractal detrended fluctuation analysis (MF-DFA) proposed by Kantelhardt et al. (2002). The usefulness of this method on complex bioelectrical signals has been demonstrated by several researchers in the past.

4.5 Electromyography Data

In our study we obtained EMG data from "Physionet" which is a database that contains different types of physiological data. Three subjects one belonging to the healthy group, one with myopathy, and another with neuropathy were chosen for the study. The details of the data are illustrated below.

Data were collected with a Medelec Synergy N2 EMG Monitoring System (Oxford Instruments Medical, Old Woking, United Kingdom). EMG data are from:

1. A 44-year-old man without history of neuromuscular disease
2. A 57-year-old man with myopathy due to long-standing history of polymyositis, treated effectively with steroids and low-dose methotrexate
3. A 62-year-old man with chronic low back pain and neuropathy due to a right L5 radiculopathy

The data were recorded at 50 KHz and then down-sampled to 4 KHz. During the recording process, two analog filters were used: a 20 Hz high-pass filter and a 5 KHz low-pass filter. The data were further divided into five equal sets for each subject. A plot of each signal for the three different groups for 1 s has been illustrated in Figs. 4.1a, 4.1b, and 4.1c.

4.6 Multifractal Detrended Fluctuation Analysis of EMG Signals

We know that multifractal detrended fluctuation analysis is a non-linear analysis technique, the application of which on a given set of data provides information about any evidence of self-similarity or persistence in the series (McArthur et al. 2013). Since the above literature provides sufficient evidence of self-similarity of EMG signals, hence we have been motivated to explore the EMG signals obtained from "Physionet" EMG database [https://physionet.org/physiobank/database/emgdb] using a self-similarity detection technique. The detail of the method is outlined in the Appendix A.

Wang et al. (2007) analyzed isometric muscle fatigue of biceps brachii using singularity spectrum of MF-DFA. Fractal features can help to classify SEMG signals of forearm muscles. In order to understand the source of multifractality in the SEMG signals obtained from healthy normal subjects in fatigue and non-fatigue circumstances, Marri and Swaminathan (2015a) transformed the signals into shuffled and surrogate series and then applied MF-DFA technique. They found correlation to be the main source of multifractality. They concluded that this multifractal analysis method can be used for characterizing the changes that occur during muscle contraction in various neuromuscular studies. Marri and Swaminathan (2015b) proposed a method for analyzing SEMG signals and identified onset of muscle fatigue using multifractal detrending moving average algorithm (MF-DMA). In another study Marri and Swaminathan (2016) applied MF-DMA to analyze SEMG data from biceps brachii muscles of healthy participants The SEMG signal identified a range of fractal exponents of different scales in fatigue and non-fatigue condition. The authors found that the scaling and generalized Hurst exponent indicate the influence of higher amplitude and lower amplitude fluctuation during fatigue condition.

In our work we used MF-DFA, the method proposed by Kantelhardt et al. (2002) to transform the data sets of EMG of healthy subject, people suffering from myopathy and neuropathy to extract the integrated signal. The method helps to minimize the noise in the time series efficiently. The integrated time series was then divided to N_s bins, where $N_s = \text{int}(N/s)$, and the fluctuation function $F_q(s)$ was determined for an index q varying between -10 and $+10$ in steps of 1. A plot of log $F_q(s)$ vs. log s for all the three sets of healthy, myopathy, and neuropathy is depicted in Figs. 4.2a, 4.2b, and 4.2c, respectively. We can see from all the plots that log $F_q(s)$ varies linearly with log s. If such a scaling exists with increase in value of s, the fluctuation function $F_q(s)$ increases and shows power–law behavior for long-range power

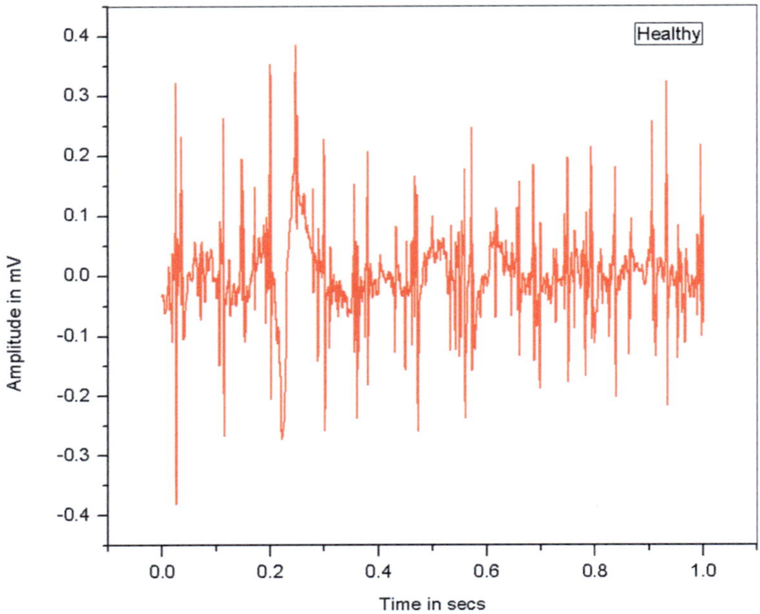

Fig. 4.1a Plot of EMG signal of a healthy subject for 1 s

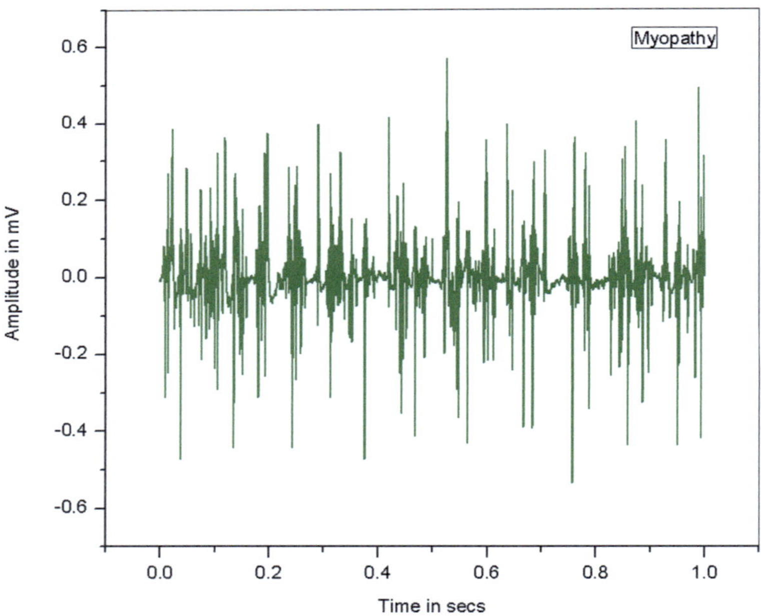

Fig. 4.1b Plot of EMG signal of a myopathy subject for 1 s

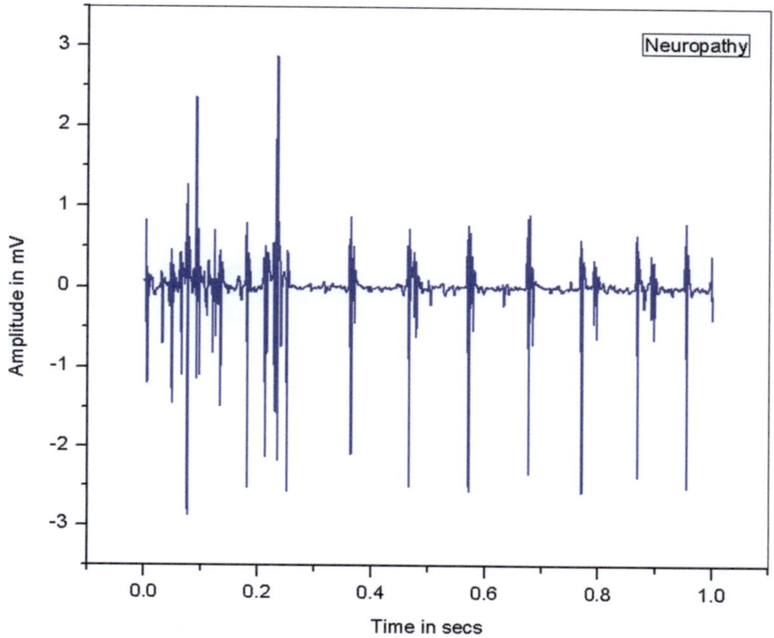

Fig. 4.1c Plot of EMG signal of a neuropathy subject for 1 s

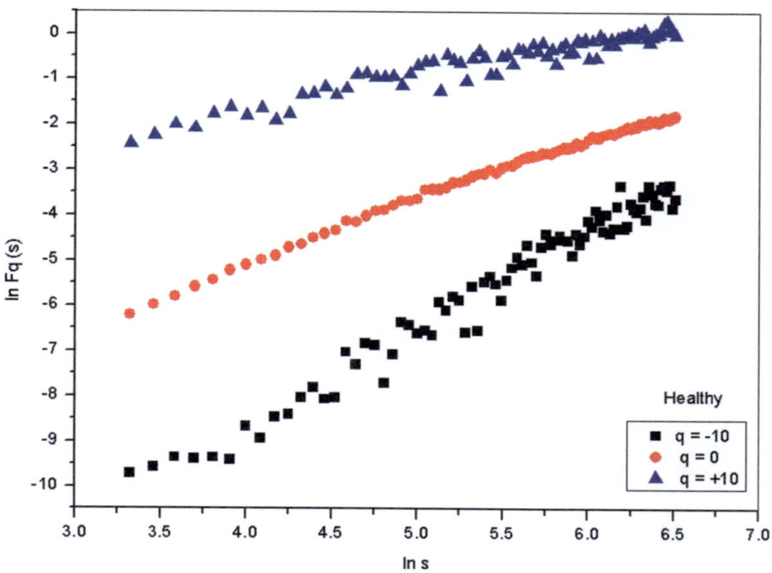

Fig. 4.2a Plot of ln F_q vs. ln s for a particular set of EMG signal of healthy subject (Ghosh et al. 2017)

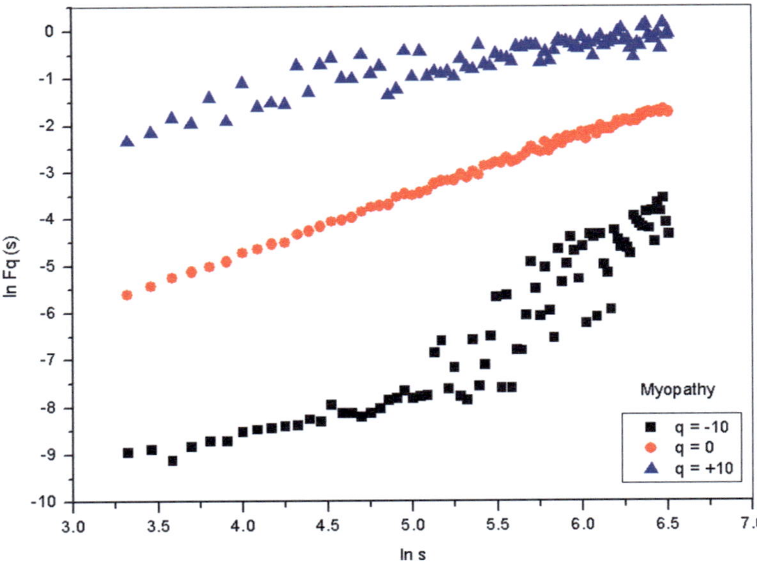

Fig. 4.2b Plot of ln F_q vs. ln s for a particular set of EMG signal of myopathy subject (Ghosh et al. 2017)

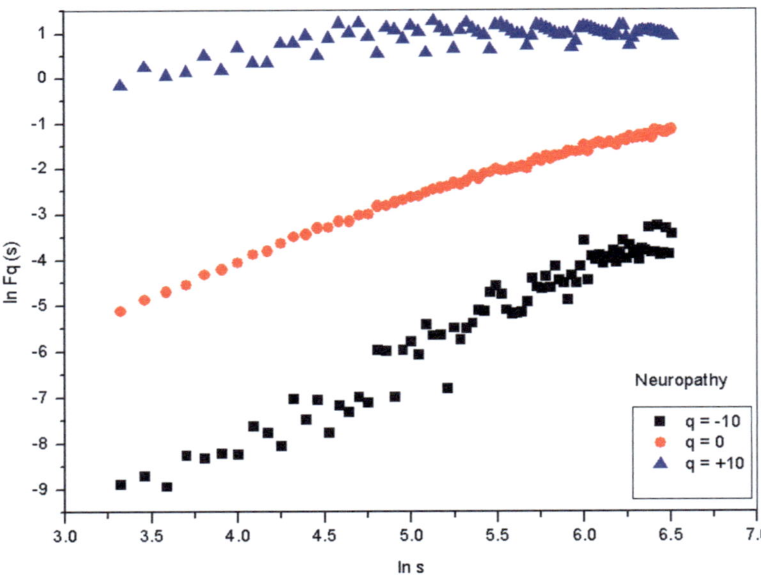

Fig. 4.2c Plot of ln F_q vs. ln s for a particular set of EMG signal of neuropathy subject (Ghosh et al. 2017)

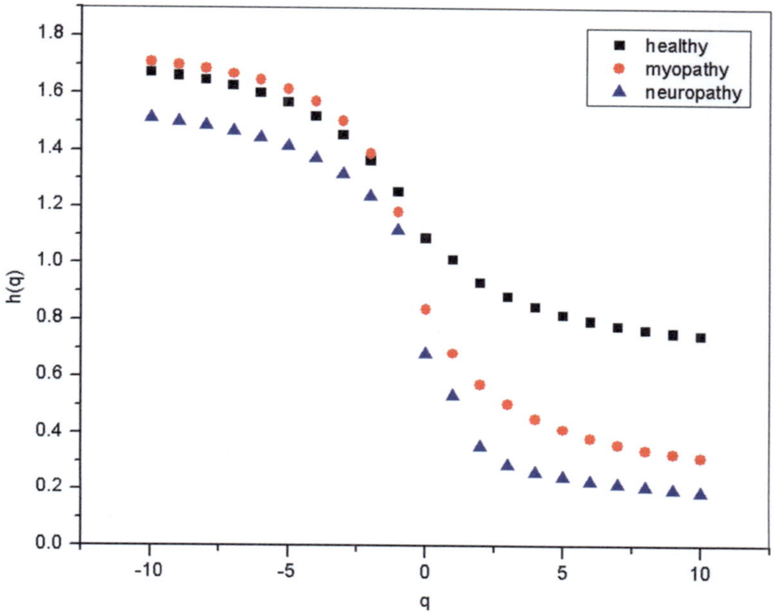

Fig. 4.3 Plot of $h(q)$ vs. q for a particular set of EMG signals of healthy, myopathy, and neuropathy (Ghosh et al. 2017)

correlated series, and then the linear dependence of the plot is an indication of scaling behavior. Thus the plot of log $F_q(s)$ vs. log s is indicative of scaling behavior for healthy, myopathy, and neuropathy subjects. From the slope of linear fit of log $F_q(s)$ vs. log s plots, we obtain the values of generalized Hurst exponent $h(q)$. We know a unique value of Hurst exponent $h(q)$ for all values of q is indicative of a monofractal time series. But here from the plot of $h(q)$ vs. q (Fig. 4.3), we can see $h(q)$ to vary with q. When Hurst exponent $h(q)$ is found to depend on q, the time series is thought to be multifractal. It has been shown by Kantelhardt et al. (2003) that for $q < 0$, the Hurst exponent value will be more than that for $q > 0$. Thus the plot of generalized Hurst exponent $h(q)$ against q proves multifractality of EMG time series of all the sets.

Next we determined the values of classical scaling exponent $\tau(q)$. For a monofractal series, the scaling exponent $\tau(q)$ is found to depend linearly with q, whereas for a multifractal series, the scaling exponent $\tau(q)$ depends non-linearly on q. Thus from Fig. 4.4 we can observe the multifractality of EMG signals of healthy, myopathy, and neuropathy as $\tau(q)$ is found to depend non-linearly on q (Ghosh et al. 2017). Since multifractal signals have multiple Hurst exponent, $\tau(q)$ is found to depend non-linearly on q (Ashkenazy et al. 2003a). Hence multifractal feature of EMG signals is evidenced from dependence of both $\tau(q)$ and $h(q)$ on q. Further degree of dependence of $h(q)$ on q can be observed from Fig. 4.3 which implies that the degree of multifractality is different in different cases (Ghosh et al. 2017).

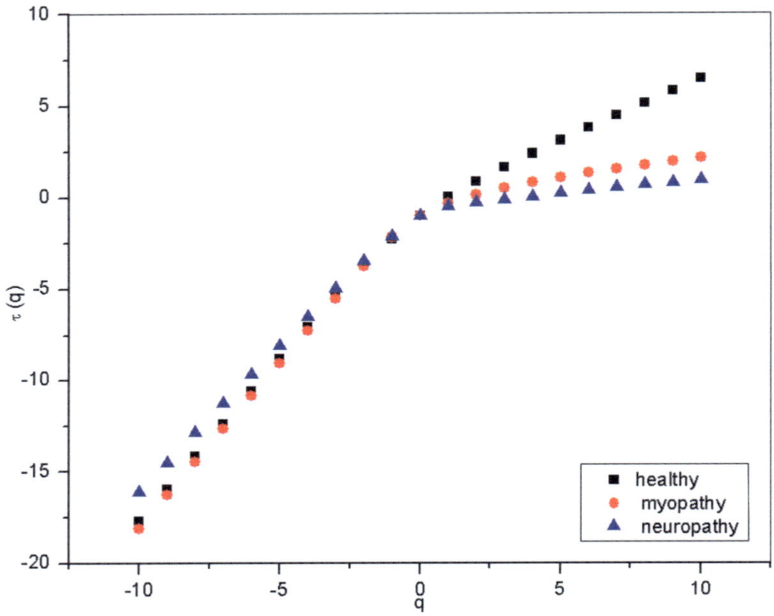

Fig. 4.4 Plot of $\tau(q)$ vs. q for a particular set of EMG signals of healthy, myopathy, and neuropathy (Ghosh et al. 2017)

Table 4.1 depicts the values of Hurst exponent $h(q)$ of healthy, myopathy, and neuropathy subjects. The Hurst exponent values are seen to decrease as q increases from -10 to $+10$. For $q = 2$ we can see the generalized Hurst exponent $h(q)$ of healthy and myopathy subjects are greater than 0.5. We know the exponent $h(q = 2)$ is equal to Hurst index (Peng et al. 1994). Value of h at $q = 2$ is important as it provides explanation about the nature of the time series, i.e., whether the series is a random process or long-range anticorrelated or correlated. Value of $h(q = 2) = 0.5$ corresponds to scale invariance of the series, and the series is an independent random process. For $h(q = 2) < 0.5$ the series is long-range anticorrelated and if $0.5 < h(q = 2) < 1$, the series shows long-term correlated behavior, which is a multifractal character. Thus from Table 4.1 we can see that for healthy and myopathy subjects, the value of generalized Hurst exponent $h(q = 2)$ is more than 0.5 which is an evidence that long-range correlation and persistent properties are present in the sets, whereas for neuropathy subject, $h(q = 2)$ is less than 0.5, which indicates anticorrelation and anti-persistence of the EMG signal (Ghosh et al. 2017).

A quantitative determination of the degree of multifractality can also be done from the multifractal spectrum. The width of the multifractal spectrum ($f(\alpha)$ vs. α) has been attributed to be the measure of the degree of multifractality (Ashkenazy et al. 2003b). Figure 4.5 shows the multifractal spectrum of healthy, myopathy, and neuropathy EMG signals. Shimizu et al. (2002) have noted that from the

Table 4.1 Values of $h(q)$ corresponding to q for a particular set of EMG signals of healthy, myopathy, and neuropathy subjects (Ghosh et al. 2017)

Order q	Generalized Hurst exponent $h(q)$		
	Healthy	Myopathy	Neuropathy
−10	1.67	1.71	1.51
−9	1.66	1.70	1.50
−8	1.65	1.69	1.49
−7	1.63	1.67	1.47
−6	1.60	1.65	1.44
−5	1.57	1.61	1.41
−4	1.52	1.57	1.37
−3	1.45	1.50	1.32
−2	1.36	1.39	1.24
−1	1.26	1.18	1.12
0	1.09	0.84	0.68
1	1.01	0.68	0.54
2	0.93	0.57	0.36
3	0.88	0.51	0.29
4	0.85	0.46	0.26
5	0.82	0.42	0.25
6	0.80	0.39	0.23
7	0.78	0.36	0.22
8	0.77	0.34	0.21
9	0.75	0.33	0.20
10	0.75	0.32	0.19

multifractal spectrum, the relative importance of various fractal exponents in the time series can be extracted, thus denoting width of the spectrum to a range of exponents (Shimizu et al. 2002).

In Table 4.2 the values of multifractal width w obtained by fitting the multifractal spectrums about the neighborhood of maximum are listed, where we can observe that the multifractal widths in five sets of all the three healthy, myopathy, and neuropathy EMG signals are different ranging from as low as 1.144 to as high as 1.257, from 1.507 to 1.605, and from 1.655 to 1.991, respectively, giving a clear indication of increasing complexity from healthy subject to neuropathy subject (Ghosh et al. 2017). We know that the auto-correlation exponent (γ) gives an estimate of the degree of correlation in a time series. Table 4.3 lists the values of auto-correlation exponent (γ) of EMG signals of all the three sets. A lower value of γ is an indication of strong correlation and a value of 1 is indicative of uncorrelated behavior. Thus from the values obtained in Table 4.3, we can see that healthy subject shows least value of auto-correlation exponent (γ), thus confirming a strong correlation, whereas myopathy patient shows values close to 1 indicating weaker correlation in EMG signal and neuropathy subjects greater than 1 implying no correlation at all. Thus these values confirm loss of complexity in case of myopathy and neuropathy subjects.

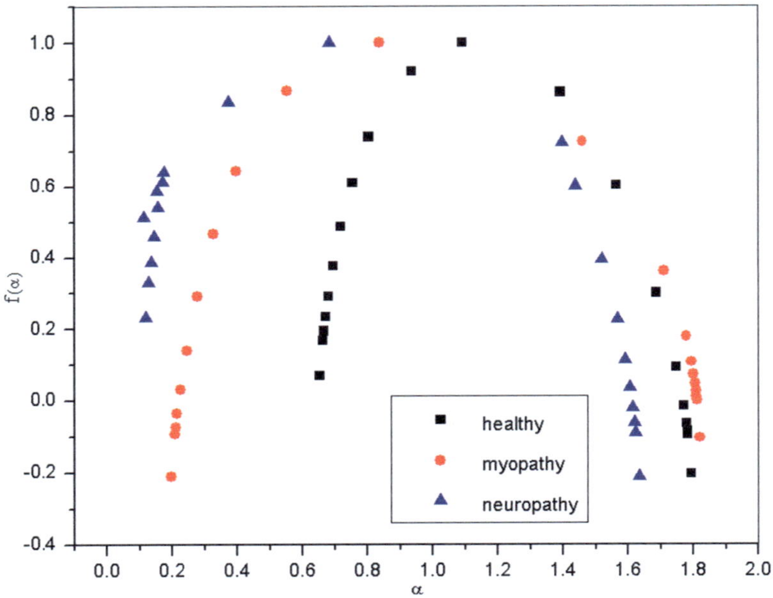

Fig. 4.5 Plot of $f(\alpha)$ vs. α for a particular set of EMG signals of healthy, myopathy, and neuropathy (Ghosh et al. 2017)

Table 4.2 Values of w for all the five sets of EMG signals of healthy, myopathy, and neuropathy subjects (Ghosh et al. 2017)

Set	Multifractal width (w)		
	Healthy	Myopathy	Neuropathy
1	1.161 ± 0.042	1.605 ± 0.078	1.655 ± 0.140
2	1.146 ± 0.041	1.583 ± 0.077	1.848 ± 0.103
3	1.257 ± 0.026	1.598 ± 0.087	1.855 ± 0.105
4	1.230 ± 0.050	1.507 ± 0.078	1.991 ± 0.078
5	1.144 ± 0.041	1.598 ± 0.073	1.813 ± 0.082

Table 4.3 Values of γ for all the five sets of EMG signals of healthy, myopathy, and neuropathy subjects (Ghosh et al. 2017)

Set	Auto-correlation exponent (γ)		
	Healthy	Myopathy	Neuropathy
1	0.132 ± 0.004	0.852 ± 0.010	1.288 ± 0.007
2	0.075 ± 0.006	0.842 ± 0.011	1.462 ± 0.010
3	0.069 ± 0.005	0.793 ± 0.009	1.442 ± 0.009
4	0.262 ± 0.004	0.73 ± 0.010	1.459 ± 0.009
5	0.035 ± 0.005	0.763 ± 0.010	1.431 ± 0.009

Only a few works have reported the multifractality of EMG signals. Earlier in the above literature, we have briefed the work of Gang et al. (2007) where they used a multifractal approach. Chhabra and Jensen (1989) showed multifractality of SEMG signals. The area of the multifractal spectrum of the SEMG signals was found to increase significantly during muscle fatigue. Thus they concluded that the area of the multifractal spectrum could then be used as an assessor of muscle fatigue which is more sensitive than the single characteristic frequency such as the median frequency (MDF) or mean frequency (MNF) of the power spectral density (PSD) which was a then popular method of estimating fatigue (Lindstrom et al. 1977; Stulen and De Luca 1981). In their opinion the large area of SEMG multifractal singularity spectrum is a reflection of the strengthened activity of the nervous system in the process of muscle fatigue (Lindstrom et al. 1977). In another work Talebinejad et al. (2009) used a bi-phase power spectrum method (BPSM) for fractal analysis of SEMG signals and also included an algorithm for extraction of fractal indicators (FIs). For force and joint angle, BSPM was evaluated. Fractal indicators demonstrated the changes that reflect in EMG signals. BSPM was compared with geometrical techniques and the $1/f^{\alpha}$ approach for fractal analysis of electromyography signals. They concluded that BPSM provides reliable information, as it consists of components which are capable of sensing force and joint angle effects separately, which could be used as complementary information for confounded conventional measures (Ghosh et al. 2017).

Though these methods demonstrate their usefulness in characterizing various EMG signals, multifractal analysis using MF-DFA technique has been established as a superior analysis technique than wavelet transform modulus maxima (WTMM) in terms of reliable applications (Oswiecimka et al. 2006). Compared to other conventional methods, MF-DFA has reached the highest precision in scaling analysis. Some other authors too have advocated the better performance of MF-DFA than other multifractal analyses methods (Kantelhardt et al. 2002; Serrano and Figliola 2009; Huang et al. 2011) as it can detect multifractality in both stationary and nonstationary time series (Ghosh et al. 2017). Thus the study confirms that EMG signal classification to assess neuromuscular disorders is possible accurately using rigorous non-linear MF-DFA technique.

4.7 Results and Possible Advanced Level Biomarker

Using MF-DFA in our analysis, the EMG signals of healthy, myopathy, and neuropathy subjects have been classified effectively with the help of two parameters: the multifractal width (w) and auto-correlation exponent (γ). Not only have we observed different degree of multifractality thus complexity of the EMG signals of healthy, myopathy, and neuropathy subjects but have also observed the significant variation in degree of auto-correlation for all the three subjects where subject with neuropathy shows no correlation at all. Marri and Swaminathan (2016) opine that modulation of muscle contraction in a certain way by neural motor drive is an outcome of correlation present in the EMG signals. Multifractality of the EMG series was believed to be mainly due to small fluctuations. Thus it was concluded that analysis of SEMG signals by MF-DFA method can be effective in understanding the dynamics of muscles in various neuromuscular disorders.

The present study proposes a novel, rigorous method of assessment of myopathy and neuropathy using EMG time series from a different perspective. The most important point is that any EMG data available may be analyzed using the method for diagnosis and prognosis of myopathy and neuropathy. Further this method has the potential for use even for early detection of motor neuron disease (Ghosh et al. 2017). Thus electromyography data coupled with this fractal analytics can be employed to develop software which eventually can easily serve as a biomarker in predicting neuromuscular disorders.

References

Acharya UR, Ng EYK, Swapna G, Michelle YSL (2011) Classification of normal, neuropathic, and myopathic electromyography signals using non-linear dynamics method. J Med Imaging Health Inform 1:375–380

Alamedine D, Khalil M, Marque C (2013) Comparison of different EHG feature selection methods for the detection of preterm labo. Comput Math Method Med 2013:585–593

Alkan A, Gunay M (2012) Identification of EMG signals using discriminant analysis and SVM classifier. Expert Syst Appl 39:44–47

Ancillao A, Galli M, Rigoldi C, Albertini G (2014) Linear correlation between fractal dimension of surface EMG signal from rectus femoris and height of vertical jump. Chaos Solitons Fractals 66:120–126

Anmuth CJ, Goldberg G, Mayer NH (1994) Fractal dimension of EMG signals recorded with surface electrodes during isometric contractions is linearly correlated with muscle activation. Muscle Nerve 17:953–954

Arjunan SP, Kumar DK (2007). Fractal theory based non-linear analysis of SEMG. In: IEEE 3rd international conference on Intelligent Sensors, Sensor Networks and Information, 3-- 6 December 2007, pp 545–548

Arjunan SP, Kumar DK (2014) Computation of fractal features based on the fractal analysis of surface electromyogram to estimate force of contraction of different muscles. Comput Methods Biomech Biomed Engin 17:210–216

Artameeyanant P, Sultornsanee S, Chamnongthai K (2016) An EMG-based feature extraction method using a normalized weight vertical visibility algorithm for myopathy and neuropathy detection. Springer Plus 5:2101

Artemiadis PK, Kyriakopoulos KJ (2007) EMG-based teleoperation of a robot arm using low-dimensional representation. In: Proceedings of IEEE/RSJ international conference on Intelligent Robots and Systems, 29 October – 2 November 2007, pp 489–495

Artemiadis PK, Kyriakopoulos KJ (2010) EMG-based control of a robot arm using low-dimensional embeddings. IEEE Trans Robot 26:393–398

Artemiadis PK, Kyriakopoulos KJ (2011) A switching regime model for the EMG-based control of a robot arm. IEEE Trans Syst Man Cybern 41:53–63

Ashkenazy Y, Havlin S, Ivanov PC, Peng CK, Frohlinde VS et al (2003a) Magnitude and sign scaling in power-law correlated time-series. Physica A 323:19–41

Ashkenazy Y, Baker DR, Gildor H, Havlin S (2003b) Non-linearity and multifractality of climate change in the past 420,000 years. Geophys Res Lett 30:2146–2149

Basmajian J, De Luca CJ (1985) Muscles alive: their functions revealed by electromyography, 5th edn. Williams & Wilkins, Baltimore

Bue BD, Merényi E, Killian JM (2013) Classification and diagnosis of myopathy from EMG signals. In: 2nd workshop on data mining for medicine and healthcare, in conjunction with the 13th SIAM international conference on Data Mining (SDM-DMMH), Austin, TX, May 2013

Chang YC, Chang S (2002) A fast estimation algorithm on the Hurst parameter of discrete-time fractional Brownian motion. IEEE Trans Signal Process 50:554–559
Chang GC, Kang WJ, Luh JJ, Cheng CK, Lai JS et al (1996) Real-time implementation of electromyogram pattern recognition as a control command of man–machine interface. Med Eng Phys 18:529–537
Chang S, Mao ST, Hu SJ, Lin WC, Cheng CL (2000) Studies of detrusorsphincter synergia and dyssynergia during micturition in rats via fractional Brownian motion. IEEE Trans Biomed Eng 47:1066–1073
Chang S, Hu SJ, Lin WC (2004) Fractal dynamics and synchronization of rhythms in urodynamics of female Wistar rats. J Neurosci Methods 139:271–279
Chang S, Li SJ, Chiang MJ, Hu SJ, Hsyu MC (2007) Fractal dimension estimation via spectral distribution function and its application to physiological signals. IEEE Trans Biomed Eng 54:1895–1898
Chen B, Wang N (2000) Determining EMG embedding and fractal dimensions and its application. In: Proceedings of the 22nd annual EMBS international conference, Chicago IL, USA, pp 1341–1344
Chen W, Wang Z, Ren X (2006) Characterization of surface EMG signals using improved approximate entropy. J Zhejiang Univ Sci B 7:844–848
Chhabra A, Jensen RV (1989) Direct determination of the f(α) singularity spectrum. Phys Rev Lett 62:1327–1330
Dang KTQ, Minh HL, Thanh HN, Van TV (2012) Analyzing surface EMG signals to determine relationship between jaw imbalance and arm strength loss. Biomed Eng Online 11:55
Diab A, El-Merhie A, El-Halabi N, Khoder L (2010) Classification of uterine EMG signals using supervised classification method. Biomed Sci Eng 3:837–842
Diab A, Hassan M, Marque C, Karlsson B (2012) Quantitative performance analysis of four methods of evaluating signal non-linearity: application to uterine EMG signals. In: Proceedings of annual international conference of the IEEE Engineering in Medicine and Biology Society, pp 1045–1048
Easwaramoorthy D, Uthayakumar R (2011) Improved generalized fractal dimensions in the discrimination between healthy and epileptic EEG signals. J Comput Sci 2:31–38
Eke A, Herman P, Kocsis L, Kozak LR (2002) Fractal characterization of complexity in temporal physiological signals. Physiol Meas 23:1–38
Falconer K (2003) Fractal geometry. Wiley, New York, p 337
Farina D, Negro F (2012) Accessing the neural drive to muscle and translation to neurorehabilitation technologies. IEEE Rev Biomed Eng 5:3–14
Farina D, Merletti R, Nazzaro M, Caruso I (2001) Effect of joint angle on EMG variables in leg and thigh muscles. IEEE Eng Med Biol Mag 20:62–71
Fele GF, Kavsek G, Novak ZA, Jager F (2008) A comparison of various linear and non-linear signal processing techniques to separate uterine EMG records of term and preterm delivery groups. Med Biol Eng Comput 46:911–922
Fox CG (1989) Empirically derived relationships between fractal dimension and power law form frequency spectra. Fractals Geophy (Part Pure Appl Geophy) 131:211–239
Fuglsang-Frederiksen A (2000) The utility of interference pattern analysis. Muscle Nerve 23:18–36
Gabriel DA, Kamen G (2009) Experimental and modeling investigation of spectral compression of biceps brachii SEMG activity with increasing force levels. J Electromyogr Kinesiol 19:437–448
Gang W, Xiao-Mei R, Lei L, Zhi-Zhong W (2007) Multifractal analysis of surface EMG signals for assessing muscle fatigue during static contractions. J Zheijang Univ Sci A 8:910–915
Gerdle B, Eriksson N (1990) The behavior of mean power frequency of the surface electromyogram in Biceps brachii with increasing force and during fatigue with special regard to electrode distance. J Electromyogr Neurophysiol 30:483–489
Ghosh D, Dutta S, Chakraborty S, Samanta S (2017) Chaos based quantitative electro-diagnostic marker for diagnosis of myopathy, neuropathy and motor neuron disease. J Neurol Neurosci 8:226

Gitiaux C, Chemaly N, Quijano-Roy S, Barnerias C, Desguerre I et al (2016) Motor neuropathy contributes to crouching in patients with Dravet syndrome. Neurology 87:277–281

Gitter JA, Czerniecki MJ (1995) Fractal analysis of electromyographic interference pattern. J Neurosci Methods 58:103–108

Goen A (2014) Classification of EMG signals for assessment of neuromuscular disorders. Int J Electron Electr Eng 2:242–248

Goge A, Chan A (2004) Investigating classification parameters for continuous myoelectrically controlled prostheses. In: Proceedings of the 28th conference of the Canadian Medical & Biological Engineering Society, pp 141–144

Gokgoz E, Subasi A (2015) Comparison of decision tree algorithms for EMG signal classification using DWT. Biomed Signal Process Control 18:138–144

Gupta V, Suryanarayanan S, Reddy NP (1997) Fractal analysis of surface EMG signals from the biceps. Int J Med Inform 45:185–192

Hassan M, Terrien J, Karlsson C (2007) Comparison between approximate entropy, correntropy and time reversibility: application to uterine electromyogram signals. Med Eng Phys 33:980–986

Hassan M, Alexandersson M, Terrien J, Muszynski C, Marque C (2012) Better pregnancy monitoring using non-linear correlation analysis of external uterine electromyography. IEEE Trans Biomed Eng 60:1160–1166

Hu X, Wang ZZ, Ren XM (2005) Classification of surface EMG signal with fractal dimension. J Zhejiang Univ Sci B 6:844–848

Huang XY, Schmitt FG, Hermand JP, Gagne Y, Lu ZM (2011) Arbitrary order Hilbert spectral analysis for time series possessing scaling statistics: comparison study with detrended fluctuation analysis and wavelet leaders. Phys Rev E 84:016208–016213

Hudgins B, Parker P, Scott RN (1993) A new strategy for multifunction myoelectric control. IEEE Trans Biomed Eng 40:82–94

Janjarasjitt S (2014) Examination of the wavelet-based approach for measuring self-similarity of epileptic electroencephalogram data. J Zhejiang Univ Sci C 15:1147–1153

Jezewski J, Horoba K, Matonia A, Wrobel J (2005) Quantitative analysis of contraction patterns in electrical activity signal of pregnant uterus as an alternative to mechanical approach. Physiol Meas 26:753–767

Kang WJ, Cheng CK, Lai JS, Shiu JR, Kuo TS (1996) A comparative analysis of various EMG pattern recognition methods. Med Eng Phys 18:390–395

Kantelhardt JW, Zschiegner SA, Koscielny BE, Havlin S, Bunde A et al (2002) Multifractal detrended fluctuation analysis of nonstationary time series. Physica A 316:87–114

Kantelhardt JW, Rybski D, Zschiegner SA, Braun P, Bunde EK et al (2003) Multifractality of river runoff and precipitation: comparison of fluctuation analysis and wavelet methods. Physica A 330:240–245

Katz M (1988) Fractals and the analysis of waveforms. Comput Biol Med 18:145–156

Khalil M, Duchene J (2007) Uterine EMG analysis: a dynamic approach for change detection and classification. IEEE Trans Biomed Eng 47:748–756

Kincaid JC (2015) Nerve conduction studies and needle EMG. In: Nerves and nerve injuries, vol 1. Elsevier, London, pp 125–145

Koike Y, Kawato M (1995) Estimation of dynamic joint torques and trajectory formation from surface electromyography signals using a neural network model. Biol Cybern 73:291–300

Kupa EJ, Roy SH, Kandarian SC, de Luca CJ (1995) Effects of muscle fiber type and size on EMG median frequency and conduction velocity. J Appl Physiol 79:23–32

Lei M, Meng G (2012) Non-linear analysis of surface EMG signals. In: Naik GR (ed) Computational intelligence in electromyography analysis – a perspective on current applications and future challenges. Intech Open, pp 119–174

Lima CAM, Coelho A, Madeo RCB, Peres SM (2016) Classification of electromyography signals using relevance vector machines and fractal dimension. Neural Comput Appl 27:791–804

Lindstrom L, Kadefors R, Petersen I (1977) An electromyographic index for localized muscle fatigue. J Appl Physiol Respir Environ Exerc Physiol 43:750–754
Lucovnik M, Maner LW, Chambliss LR, Blumrick R, Balducci R et al (2011) Noninvasive uterine electromyography for prediction of preterm delivery. Am J Obstet Gynecol 204:228–2e1
Marri K, Swaminathan R (2015a) Analyzing origin of multifractality of surface electromyography signals in dynamic contractions. J Nanotechnol Eng Med 6:031002–031001
Marri K, Swaminathan R (2015b) Identification of onset of fatigue in biceps Brachii muscles using surface EMG and multifractal DMA alogrithm. Biomed Sci Instrum 51:107–114
Marri K, Swaminathan R (2016) Analysis of biceps Brachii muscles in dynamic contraction using sEMG signals and multifractal DMA algorithm. Int J Signal Process Syst 4:79–85
McArthur L, Mackenzie S, Boland J (2013) Multifractal analysis of wind farm power output. In: 20th international Congress on Modeling and Simulation (MODSIM 2013), Adelaide, Australia, 1–6 December 2013, pp 420–426
Mishra VK, Bajaj V, Kumar A, Singh GK (2016) Analysis of ALS and normal EMG signals based on empirical mode decomposition. IET Sci Measurement Technol 10:963–971
Monsifrot J, Carpentier EL, Aoustin Y (2004) Sequential decoding of intramuscular EMG signals via estimation of a Markov model. IEEE Trans Neural Syst Rehabil Eng 22:1030–1038
Moreside JM, Quirk DA, Hubley-Kozey CL (2014) Temporal patterns of the trunk muscles remain altered in a low back-injured population despite subjective reports of recovery. Arch Phys Med Rehabil 95:686–698
Naeem SM, Seddik AF, Eldosoky MA (2014) New technique based on uterine electromyography non-linearity for preterm delivery detection. J Eng Technol Res 6:107–114
Naik G, Kumar D, Arjunan S (2009) Use of SEMG in identification of low level muscle activities: features based on ICA and fractal dimension. Conf Proc IEEE Eng Med Biol Soc 2009:364–367
Naik GR, Selvan SE, Nguyen HT (2016) Single-channel EMG classification with ensemble-empirical-mode-decomposition- based ICA for diagnosing neuromuscular disorders. IEEE Trans Neural Syst Rehabil Eng 24:734–743
Najarian K, Splinter R (2012) Biomedical signal and image processing, 2nd edn. CRC Press/Taylor & Francis Group, Boca Raton/London/New York
Nikolic M, Krarup C (2011) EMGTools, an adaptive and versatile tool for detailed EMG analysis. IEEE Trans Biomed Eng 58:2707–2718
Nussbaum MA, Yassierli (2003) Assessment of localized muscle fatigue furing low-moderate static contractions using the fractal dimension of EMG. In: Proceedings of the XVth triennial Congress of the International Ergonomics Association, Seoul, Korea, August 25–29
Oswiecimka P, Kwapien J, Drozdz S (2006) Wavelet versus detrended fluctuation analysis of multifractal structures. Phys Rev E 74:06103–06137
Patidar M, Jain N, Parikh A (2013) Classification of normal and myopathy EMG signals using BP neural network. Int J Comput Appl 69:0975–8887
Peng CK, Buldyrev SV, Havlin S, Simons M, Stanley HE et al (1994) Mosaic organization of DNA nucleotides. Phys Rev E 49:1685–1689
Phinyomark A, Phothisonothai M, Limsakul C, Phukpattaranont P (2009) Detrended fluctuation analysis of electromyography signal to identify hand movement. In: Proceedings of the second Biomedical Engineering international conference, Phuket, Thailand, August 13–14, 2009, pp 324–329
Phinyomark A, Phothisonothai M, Limsakul C, Phukpattaranont P (2010) Effect of trends on detrended fluctuation analysis for surface electromyography (EMG) signal. In: The eighth PSU Engineering conference 22–23 April 2010, pp 333–338
Phinyomark A, Phukpattaranont P, Limsakul C (2012) Fractal analysis features for weak and single-channel upper-limb EMG signals. Expert Syst Appl 39:11156–11163
Ravier P, Buttelli O, Jennane R, Couratier P (2005) An EMG. Fractal indicator having different sensitivities to changes in force and muscle fatigue during voluntary static muscle contractions. J Electromyogr Kinesiol 15:210–221

Ren P, Yao S, Li J, Valdes-Sosa PA, Kendrick KM (2015) Improved prediction of preterm delivery using empirical mode decomposition analysis of uterine electromyography signals. PLoS One 10:e0132116

Riillo F, Quitadamo L, Cavrinia F, Gruppioni E, Pinto C et al (2014) Optimization of EMG-based hand gesture recognition: supervised vs. unsupervised data preprocessing on healthy subjects and transradial amputees. Biomed Signal Process Control 14:117–125

Ryu W, Han B, Kim J (2008) Continuous position control of 1 DOF manipulator using EMG signals. In: Proceeding in 3rd international conference on Convergence and Hybrid Information Technology, 11–13 November, 2008, pp 870–874

Sarkar M, Leong TY (2003) Characterization of medical time series using fuzzy similarity-based fractal dimensions. Artif Intell Med 27:201–222

Serrano E, Figliola A (2009) Wavelet leaders: a new method to estimate the multifractal singularity spectra. Physica A 388:2793–2805

Sharma RR, Chandra P, Pachori RB (2017) Electromyogram signal analysis using eigenvalue decomposition of the Hankel Matrix. In: International conference on machine intelligence and signal processing at: Indian Institute of Technology Indore, Indore, India, November, 2017

Shields R (2006) Fractal dimension of the EMG interference pattern: preliminary observations and comparisons with other measures of interference pattern analysis. J Clin Neurophysiol 10:117–118

Shimizu Y, Thurner S, Ehrenberger K (2002) Multifractal spectra as a measure of complexity in human posture. Fractals 10:103–116

Smith RJ, Tenore F, Huberdeau D, Etienne-Cummings R, Thakor NV (2008) Continuous decoding of finger position from surface EMG signals for the control of powered prostheses. In: Proceedings of IEEE 30th annual international Conference on Engineering in Medicine and Biology Society, August 2008, pp 197–200

Stulen FB, De Luca CJ (1981) Frequency parameters of the myoelectric signal as a measure of muscle conduction velocity. IEEE Trans Biomed Eng 28:515–523

Subasi A (2013) Classification of EMG signals using PSO optimized SVM for diagnosis of neuromuscular disorders. Comput Biol Med 43:576–586

Talebinejad M, Chan ADC, Miri A, Dansereau RM (2009) Fractal analysis of surface electromyography signals: a novel power spectrum-based method. J Electromyogr Kinesiol 19:840–850

Trojaborg W (1987) Motor unit disorders and myopathies. In: Halliday MA, Butler RJ, Paul R (eds) A textbook book of clinical neurophysiology. Wiley, New York, pp 417–438

Vishnu RS, Shalu GK (2015) Identification of surface EMG – angular velocity model using artificial neural network. Int J Adv Res Electr Electron Instrument Eng 4:7201–7208

Wang G, Ren X, Li L, Wang Z (2007) Multifractal analysis of surface EMG signals for assessing muscle fatigue during static contractions. J Zheijang Univ Sci A 8:910–915

Webber CL Jr, Schmidt MA, Walsh JM (1995) Influence of isometric loading on biceps EMG dynamics as assessed by linear and non-linear tools. J Appl Physiol 78:814–822

Weiss JM, Weiss LD, Silver JK (2015) Neuromuscular junction disorders, easy EMG: a guide to performing nerve conduction studies and electromyography. Elsevier, London. ISBN:978-0-323-28664-0

Xu Z, Xiao S (1997) Fractal dimension of surface EMG and its determinants. In: Proceedings of 19th international conference – IEEE/EMBS, Chicago, IL, USA, pp 1570–1573

Zhao J, Jiang L, Cai H, Liu H (2007) EMG pattern recognition method for prosthetic hand based on wavelet transform and sample entropy. Control Decis 22:927–930

Chapter 5
Multifractal Study of Parkinson's and Huntington's Diseases with Human Gait Data

Abstract In this chapter we have presented how multifractal detrended cross-correlation analysis technique can be used to study Parkinson's disease from human gait pattern of those patients when compared to those of normal people. The chapter further emphasizes that this study is important as a new novel technique whereby data from the correlation between the two feet provides status of the degree of neurodegenerative disorder. The chapter also presents how multifractal methodologies can also be applied in Huntington's disease.

5.1 Introduction

Locomotion is defined as the pattern of movement of animals from one geographic location to another. From stating to stopping locomotion has many stages. These stages which also includes change in speed and direction together combine to produce displacement of body parts in specified manner that helps animals to move forward (Kuo 2002; Inman et al. 2006). The human locomotion mechanism is controlled by the central nervous system in coordination with the musculoskeletal system. Premature motor control in young children generates walking patterns which is not stable which leads to unpredictable posture. As children first learn to walk, huge fluctuations are noted from one stride interval to the next (Hillmana et al. 2009). As motor skills develop, stride variability decreases (Hausdorff et al. 1999; Holt et al. 2007) from childhood to adulthood.

A strong connection has been demonstrated between human walking and random walk by Hausdorff et al. (2001). Although walking appears to us a regular periodic process, even under stationary conditions, the gait pattern reveals small fluctuations. Gait is basically the pattern of movement of limbs. Human locomotion can be described by three distinct stages: (1) development stage (from resting position to some velocity), (2) rhythmic stage (at some constant velocity), and (3) decay stage (back to the rest position) (Scafetta et al. 2009). Human step has two different phases. With the striking of ground, the first phase commences which ends when the foot is lifted, and in the second phase the foot is lifted, and when it strikes the ground again, the phase is completed (Singh et al. 2013). Human locomotion is not

D. Ghosh et al., *Multifractals and Chronic Diseases of the Central Nervous System*,
https://doi.org/10.1007/978-981-13-3552-5_5

only a voluntary but is also an automatic operation whose rhythmic movements are controlled by both feed forward and feedback control (Kuo 2002). Central pattern generator (CPG) is comprised of an arrangement of firing neurons that can create a synchronized output in the form of movement of muscles in a given succession (Collins and Richmond 1994; Winters and Crago 2000). Because of identical cyclic repetition of the movement of each limb (Griffin et al. 2000), its fluctuation can be defined by a non-linear oscillator for participation of each limb in the locomotion mechanism (Collins and Stewart 1993).

Conventional linear measure like standard deviation which is used to explore the dynamics of human movement variability provides a measure of the extent or magnitude of the variability about a central point. Lomax (2007) was of the opinion that conventional linear methods used to examine variability presume that changes that occur in between repetition of a task are random and independent from a statistical viewpoint. Some previous studies have distinguished these variations from noise (Dingwell and Cusumano 2000; Stergiou et al. 2004; Dingwell and Kang 2007; Delignières and Torre 2009), while some other studies have indicated the deterministic origin of these variations (Dingwell and Cusumano 2000; Miller et al. 2006; Dingwell and Kang 2007; Harbourne and Stergiou 2009). Thus they are neither random nor independent. Though stride variations during walking may apparently appear to be random without any connection between the present and future strides, but the healthy adult locomotor system is well connected where the fluctuations from one stride to the next exhibit a sophisticated, hidden temporal structure. Thus mathematical techniques in the domain of non-linear dynamics have been developed to provide proper estimate of the temporal structure of variability in contrast to traditional statistical tools which only determine the magnitude of variation (Stergiou and Decker 2011).

Several studies have considered human gait to be a complex, non-linear process by which the locomotor system incorporates input from the cerebellum, the motor cortex, and the basal ganglia, as well as feedback from visual, vestibular, and proprioceptive sensors. Under healthy conditions, the locomotor system produces a stable walking pattern; the kinetics, kinematics, and muscular activity of gait remain relatively constant from one step to the next, even during unconstrained walking (Inman et al. 1981; Winter 1984; Palta 1985; Kadaba et al. 1989; Pailhous and Bonnard 1992; Decker et al. 2010). This is the reason why most conventional biomechanical studies are based on the thorough analysis of a walking cycle. The data obtained are extrapolated into the whole walking process. Studies using non-linear dynamics have revealed fluctuations of gait patterns even under apparently stable conditions (Guimares and Isaacs 1980; Gabell and Nayak 1984; Yamasaki et al. 1984, 1991). Stride interval fluctuations have been found to exhibit fractal properties and long-range correlations in healthy, young adults (Hausdorff 2007). Thus, many studies have tried to elucidate (Van Emmerik et al. 2004) the complex behavior human gait dynamics using practical applications mainly focusing on aging and pathologies affecting human walking (Torres et al. 2013). Study of human gait for normal and diseased set using different methods has also been reported (Hausdorf et al. 1995, 1996, 1997, 2000; Goldberger et al. 2002a, b; Scafetta et al. 2003, 2007; Van Orden et al. 2009).

Several devices have analyzed human gait complexity using pressure or force sensors (Webster et al. 2005; Beauchet et al. 2008) or hip and knee angles (Davis et al. 1991). Mathematical methods have too been useful in quantifying the complexity of gait dynamics. Costa et al. (2003) used multiscale entropy method to compare the difference in complexity of gait in between walking speeds. With the use of an entropy-based method, Kurz and Stergiou (2003) showed that aging may cause neurophysiological changes which may affect in selecting a stable gait with certainty. Matjaz (2005) used the largest Lyapunov exponent to determine the dynamics of human gait. In a group of patients suffering from anterior cruciate ligament deficit, Moraiti et al. (2007) employed Lyapunov exponent (LyE) and found that the group presented more rigid and predictable walking patterns compared to control group. The study suggests diminished system complexity and reduced functional response. Other works using LyE have also been reported to characterize the underlying gait complexity during movement (Stergiou et al. 2004; Buzzi and Ulrich 2004; Yoshino et al. 2004; Kurz and Stergiou 2007). To predict everyday walking activities, Cavanaugh et al. (2007) applied approximate entropy (ApEn) on pedometer data and reported that, in comparison to active elderly individuals, inactive individuals who walked less had more predictable walking activity. Khandoker et al. (2008) used ApEn to study the risk for falls in the elderly by analyzing the gait variability. Smith et al. (2011) used ApEn to estimate damaged neuromotor control of movements in early stages of life. Scafetta et al. (2009) reported fractal and multifractal properties of human gait stride intervals under different conditions. Records obtained from subjects walking at normal, slow, and fast pace speed were analyzed to determine changes in the fractal scaling as a function of the stress condition of the system. They also analyzed subjects with different ages from children to elderly and patients suffering from neurodegenerative disease to determine changes in the fractal scalings as a function of the physical maturation or degeneration of the system. They developed a supercentral pattern generator (SCPG) model that correctly prognosticates that the decrease in average of the long correlation of the stride interval time series for children and for the elderly or for those with neurodegenerative diseases can be understood as a decrease in the correlation length among the neurons of the biomechanical motor control system (MCS) due to neural maturation and neurodegeneration, respectively (Dutta et al. 2016). Tochigi et al. (2012) applied SampEn to analyze the cycle-to-cycle variability in leg acceleration signals during walking in elderly subjects and in adults with symptomatic knee osteoarthritis and found lower variability in the knee osteoarthritis compared to control group. Robbins et al. (2013) made use of principal component analysis (PCA) a multivariate statistical technique to diagnose kinematic and kinetic gait pattern (Robbins et al. 2013).

To determine stride interval fluctuations of walking of young healthy men on ground around a circular path, some researchers have used spectral analysis and detrended fluctuation analysis (DFA) (Goldberger et al. 2002b; Hausdorff et al.

1995). DFA was used in another study to characterize human gait rhythm under different walking rates (Hausdorff et al. 1996). Some works have reported the scaling exponent α to quantify walking stability (Herman et al. 2005), gait speed (Jordan et al. 2006, 2007a), and gait maturation (Hausdorff et al. 1999). Fractal-like fluctuations have also been reported in case of healthy participants running or walking on a treadmill (Frenkel-Toledo et al. 2005a; Jordan et al. 2007b).

Thus several models that have developed over the decades have helped us to understand the mechanism behind the dynamics of walking and its alterations. These models reinforce that neural mechanisms are interpretative of long-range correlations in gait dynamics. Neural mechanism changes can classify different gait alteration dynamics regardless of any changes in mechanics or peripheral function (Hausdorff 2007).

The dynamics of human gait is believed to be altered as a person ages and also with neurodegenerative diseases. With increase in age, balance and gait are affected showing decline in strength, muscle mass, and bone density. Changes in central nervous system due to age include shrinking neural soma and processes of the central cortex. With increase in the complexity of the neurologic impairment, gait variability seems to be affected (Singh et al. 2013). Neurodegenerative diseases like Parkinson's disease (PD), Huntington's disease (HD), and amyotrophic lateral sclerosis (ALS) affect the human gait dynamics. In the following sections a detailed illustration of the gait complexity in neurodegenerative patients is outlined.

5.2 Parkinson's Disease and Gait Data

Loss of dopamine-generating neurons in the basal ganglia causes Parkinson's disease (PD) which is an intensifying degenerative disorder of the central nervous system. Defective locomotion accompanied with rest tremor, bradykinesia, rigidity, and postural instability are the typical symptoms of PD (Hausdorff 2007). Fatigue, festination, small shuffling steps, and decrease in both arm swing and walking speed are some other signs of the disease (Wu and Krishnan 2010). Some works report that decrease in length of the stride and gait speed which are seen often in PD are due to bradykinetic manifestations (Morris et al. 1994a, 1996a). Parkinson's patients are not able to produce a steady gait rhythm, rather enhanced stride-to-stride variability and reduced fractal scaling index are observed in PD (Blin et al. 1990; Hausdorff et al. 1998, 2000; Frenkel-Toledo et al. 2005a). Not only patients in advanced stage of PD ((Blin et al. 1990; Hausdorff et al. 1998; Stolze et al. 2001) show enhanced stride-to-stride variability, both in the stride length and time but also patients in their early stage and who have not been administered with antiparkinsonian medications (Hausdorff et al. 1995; Baltadjieva et al. 2006).

To quantify kinetic, spatiotemporal, power spectral, and fractal parameters of the gait in PD, several researchers have used computer-aided analysis technique (Blin et al. 1990; Morris et al. 1999; Hausdorff et al. 1995, 1996, 1998, Hausdorf and Alexander 2005; Sekine et al. 2002, 2004). Morris et al. (1994b) noticed that PD patients are able to regulate steps per minute. They also observed averaged stride length to be significantly lower than healthy controls, consistent with some other works (Morris et al. 1996b, 2001; Frenkel-Toledo et al. 2005b). Acceleration signals of Parkinson's patients while climbing stairs and walking along a corridor were studied by Sekine et al. (2002) using wavelet-based fractal analysis and time-frequency matching algorithm. In another work Sekine et al. (2004) noted that acceleration signals of Parkinson's subjects obtained during one gait cycle to the next will change in a complex fashion with higher fractal dimensions of the body motion. Some other studies also suggest the fluctuation dynamics of stride interval to have increased notably in PD patients (Blin et al. 1990; Hausdorf et al. 1998, 2007). Estimating the stride-to-stride fluctuations in healthy controls and in PD patients, Hausdorff et al. (1998, 2007) obtained increased coefficient of variation in PD which is thought to be associated to severeness of the disease (Hausdorff et al. 1998). They also studied the reaction of external cueing with rhythmic auditory stimulation (e.g., by means of a metronome) on gait variability. Improved mobility and reduced fall risk in PD patients are manifested by rhythmic auditory stimulation set to 110% of the step rate (Hausdorff et al. 2007). Miller et al. (1996) reported increase of electromyographic signal variability of gastrocnemius in PD patients. To estimate the probability density functions (PDFs) of stride interval and its two sub-phases (swing interval and stance interval), for healthy subjects and PD patients, Wu and Krishnan (2010) used the Parzen-window method.

Ashkenazy et al. (2002) presented a stochastic model of gait rhythm on the basis of transitions that occur between different neural centers which replicate distinguishing statistical properties of normal human walking. The model reported changes in the dynamics of gait from childhood to adulthood with reduced correlation and volatility exponents with maturity. The model also produced time series with multifractal spectrum. Increase in volatility exponent as a function of width of the multifractal spectrum indicating change in multifractality with maturation was also observed. West and Scafetta (2003) developed a supercentral pattern generator (SCPG) model that produced both fractal and multifractal properties of gait dynamics. Using SCPG technique Scafetta et al. (2009) found human stride interval to be a complex time series that exhibits fractal and multifractal properties. The randomness of the fluctuations in elderly or subjects with neurodegenerative diseases was found to be higher.

With a motive to get a clear perception of gait dynamics of PD, we chose to quantitatively assess the degree of multifractality and cross-correlation between the total force under the left and right foot of human gait rhythm among the diseased and control set using a multifractal cross-correlation technique proposed by Zhou (2008).

5.2.1 Gait Data

PD is a chronic and progressive neurological disorder that results in tremor, rigidity, slowness, and postural instability. A disturbed gait is a common, debilitating

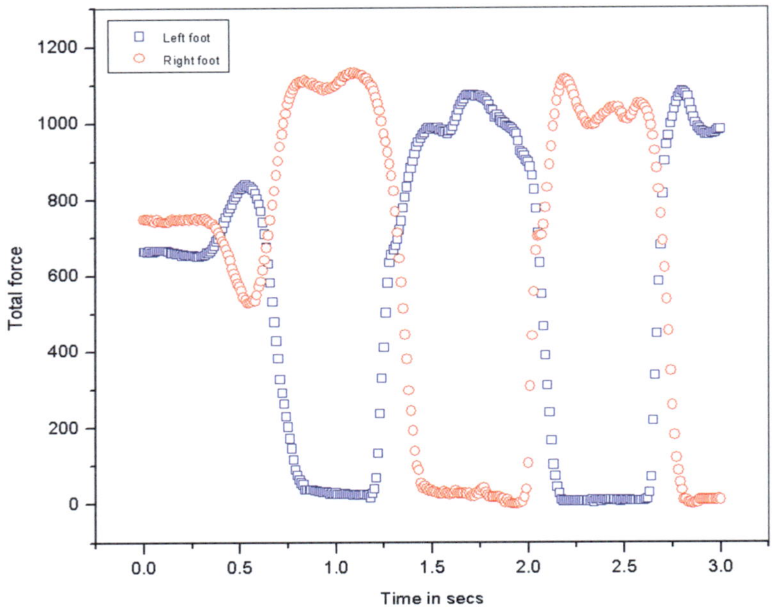

Fig. 5.1a Plot of the signal (force under each foot) for a particular Control subject for Experiment 1 for 3 s

symptom; patients with severe gait disturbances are prone to falls and may lose their functional independence. To quantify the degree of cross-correlation between the left and right foot in case of normal persons (control group) and patients with Parkinson's disease (PD), we studied the databases of two experiments from the website www.physionet.org (Yogev et al. 2005; Hausdorf et al. 2007). In the first experiment, 30 patients with idiopathic PD were compared to 28 control subjects of similar age (Yogev et al. 2005), and in the second experiment, 29 patients with idiopathic PD were compared to 26 healthy age-matched control subjects (Hausdorf et al. 2007).

The database includes the vertical ground reaction force records of subjects as they walked at their usual, self-selected pace for approximately 2 min on level ground. Underneath each foot were eight sensors (Ultraflex Computer Dyno Graphy, Infotronic Inc.) that measure force (in Newtons) as a function of time. The output of each of these 16 sensors was digitized and recorded at 100 samples per second, and the records also include 2 signals that reflect the sum of the 8 sensor outputs for each foot. For this study we have selected the data that reflect the total force under the left foot and total force under the right foot. The plot of one signal (for 3 s) from each set of experiment for both control and diseased group is shown in Figs. 5.1a, 5.1b, 5.1c and 5.1d.

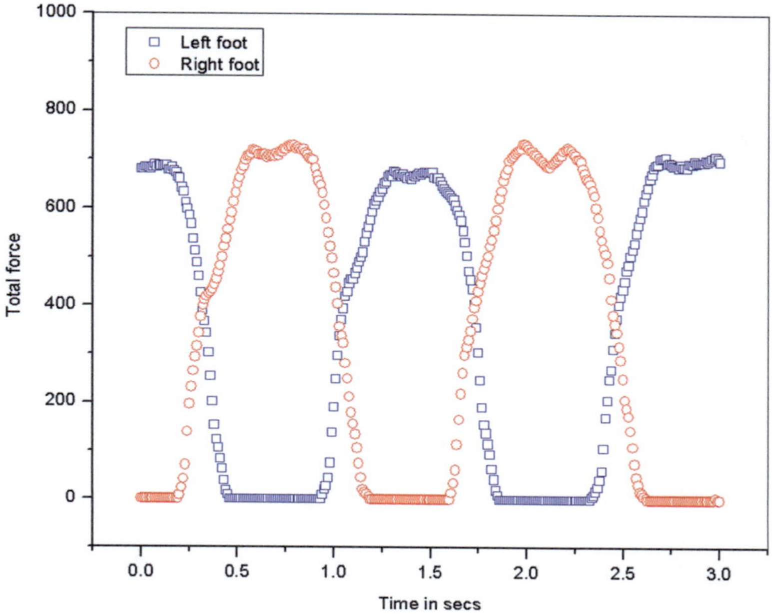

Fig. 5.1b Plot of the signal (force under each foot) for a particular Parkinson subject for Experiment 1 for 3 s

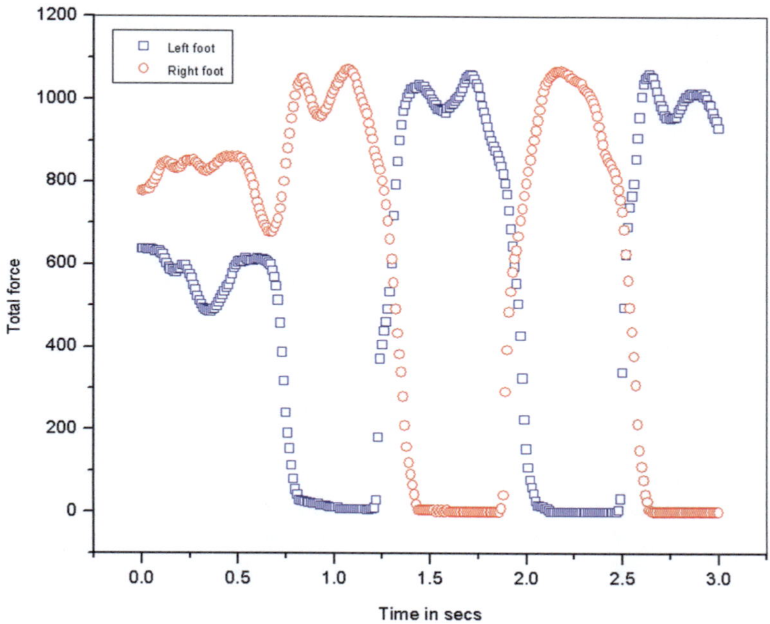

Fig. 5.1c Plot of the signal (force under each foot) for a particular Control subject for Experiment 2 for 3 s

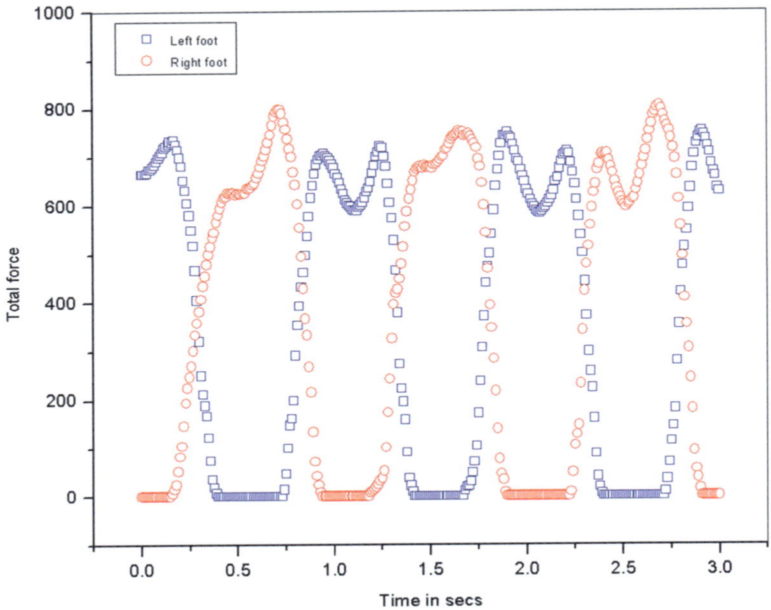

Fig. 5.1d Plot of the signal (force under each foot) for a particular Parkinson subject for Experiment 2 for 3 s

5.3 Multifractal and Multifractal Cross-Correlation Analysis of Parkinson's Disease

Monofractal or multifractal behavior may be reported while investigating a physiological time series. Owing to same scaling behavior of monofractals for the signal overall, they are indexed by a lone global exponent h_0 (the Hurst exponent) or by a single fractal dimension (Hurst 1951, Yu and Wang 2001), which indicates their stationary behavior in view of their local scaling properties. We know that compared to monofractals, multifractals are more complex as they can be dissolved into subsets and have different local Hurst exponents (h) or different fractal dimensions. This can assess the local singular behavior and describe the local scaling of the time series (Muñoz-Diosdado 2005). Due to the inherently complex and inhomogeneous character of multifractals, various exponents are required to define their scaling properties (Malamud and Turcotte 1999; Bunde et al. 2002). The relevance of multifractal formalism demonstrated by Ivanov et al. (1999) in human interbeat time series has been elaborated earlier. West and Scafetta (2003) demonstrated that gait time series, rather than being monofractal, are weakly multifractal. Muñoz-Diosdado (2005) studied the multifractal properties of stride interval time series of control subjects and patients with neurodegenerative disease. They observed narrow multifractal spectra assuming monofractal behavior for gait time series of healthy young subjects and wider spectra for old and diseased subjects. Hausdorff et al. (2001) extended the detrending technique designed for monofractal series to multifractal formalism (multifractal detrended fluctuation analysis) MF-DFA. Previous studies have emphasized the long-range correlation properties of gait series. Since in neurodegenerative gait diseases the correlation between the two feet is expected to be hampered, so instead of studying the time series of the left foot and right foot individually, a study of cross-correlation between the two feet for the diseased subjects (as compared to control group) would provide a better insight into the dynamics of gait. We have used multifractal detrended cross-correlation analysis (MF-DXA) technique since it has been widely tested in various kinds of systems and has produced effective results. The mathematical detail of the method is briefed in Appendix B.

Multifractal analysis was applied to both control and diseased (Parkinson's) set. After transforming the data sets to the integrated time series, they were divided into N_s bins where $N_s = \text{int}(N/s)$, N being the length of the series and s the length of the bin. For both the experiments the range of s was taken from 15 to $N/10$ in steps of 1. The fluctuation function $F_q(s)$ for values of q from -10 to $+10$ in steps of 1 was first determined. Plot of log F_q versus log s for a particular subject for control group and PD patients is shown in Figs. 5.2a, 5.2b, 5.2c, 5.2d, 5.2e and 5.2f. We can notice linear dependence of log F_q on log s for different values of q which suggests that there exists power–law cross-correlation between the left foot and right foot for both

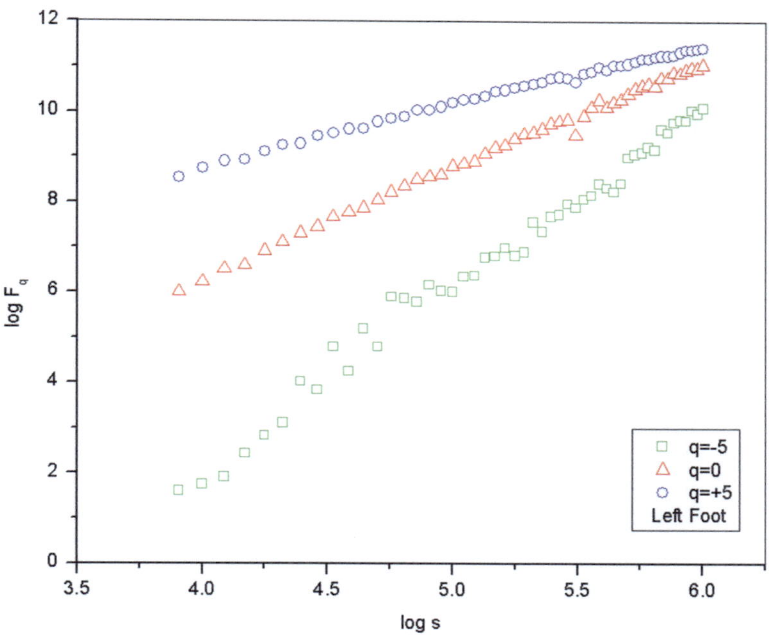

Fig. 5.2a Plot of log F_q vs. log s ($q = -5$, 0, +5) of Left Foot for a particular subject of Control group for Experiment 2 (Dutta et al. 2016)

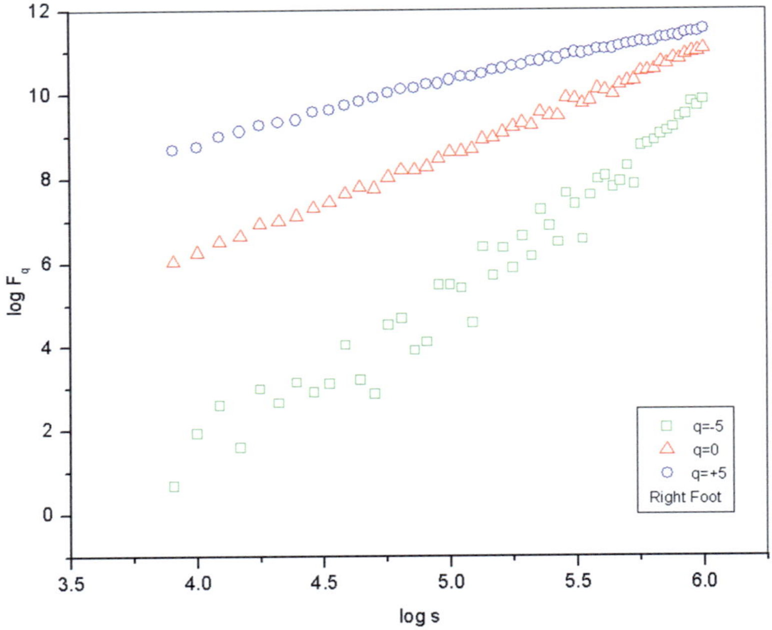

Fig. 5.2b Plot of log F_q vs. log s ($q = -5, 0, +5$) of Right Foot for a particular subject of Control group for Experiment 2 (Dutta et al. 2016)

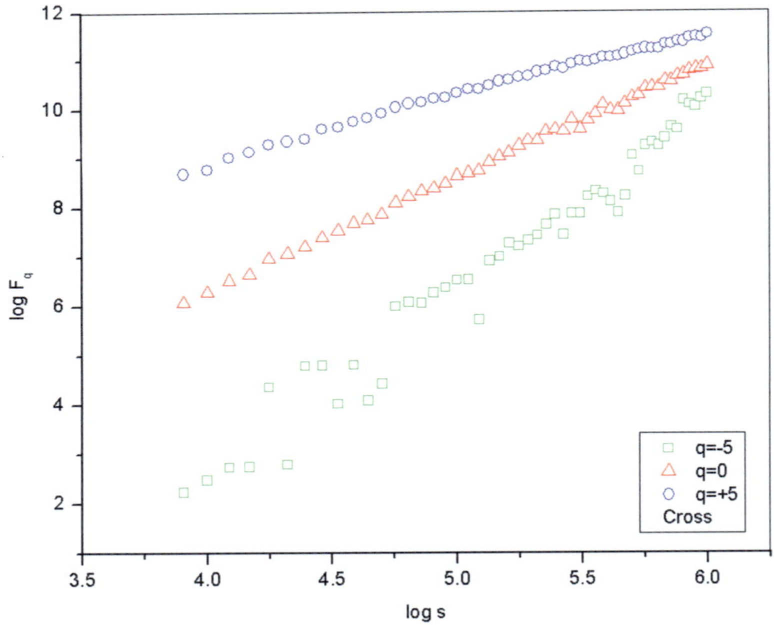

Fig. 5.2c Plot of log F_q vs. log s ($q = -5, 0, +5$) of Cross between Left and Right Foot for a particular subject of Control group for Experiment 2 (Dutta et al. 2016)

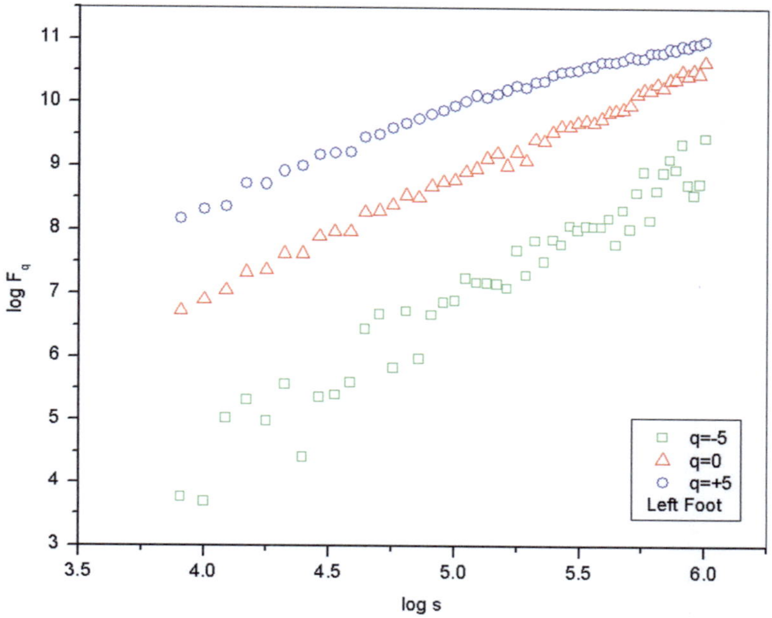

Fig. 5.2d Plot of log F_q vs. log s ($q = -5, 0, +5$) of Left Foot for a particular subject of Parkinson's group for Experiment 2 (Dutta et al. 2016)

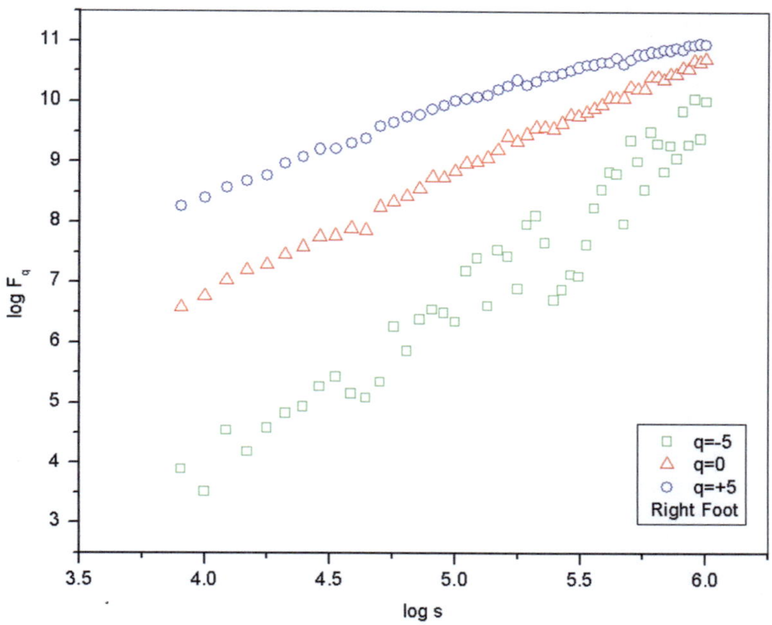

Fig. 5.2e Plot of log F_q vs. log s ($q = -5, 0, +5$) of Right Foot for a particular subject of Parkinson's group for Experiment 2 (Dutta et al. 2016)

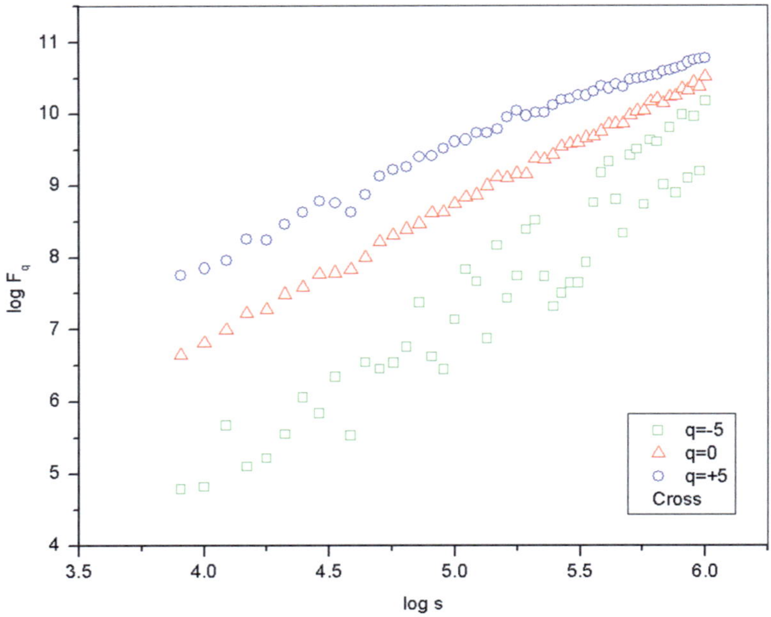

Fig. 5.2f Plot of log F_q vs. log s ($q = -5$, 0, +5) of Cross between Left and Right Foot for a particular subject of Parkinson's group for Experiment 2 (Dutta et al. 2016)

healthy persons (control group) and persons with PD. The slope of linear fit to log $F_q(s)$ vs. log s plots gives the values of Hurst exponent $h(q)$ and cross-correlation scaling exponent $\lambda(q)$. The relationship between scaling exponent $\lambda(q)$ and q for a particular healthy subject (control group) and a subject with PD for Experiments 1 and 2 are portrayed in Figs. 5.3a, 5.3b, 5.4a and 5.4b, respectively. For comparison, the generalized Hurst exponent $h(q)$ obtained from MF-DFA is also shown in the same figure. Dependence of both $h(q)$ and $\lambda(q)$ on q can be seen which advocate multifractal behavior, i.e., different power–law auto- and cross-correlations exist in the gait times series of healthy (control) and diseased (PD) subjects (Dutta et al. 2016).

The singularity spectrum $f(\alpha)$ is generalized to two cross-correlated series. The distribution of degree of cross-correlation in varied time scales can be interpreted from the singularity spectrum. To determine multifractality of cross-correlation between two series, a relation via a $\lambda(q)$ Legendre transform (Feder 1988; Peitgen et al. 1992) is obtained (detailed in Appendix B). A plot of $f(\alpha)$ vs. α is a measure of the multifractal spectrum where α gives the singularity strength or Hölder exponent and $f(\alpha)$ indicates the dimension of the subset of the series characterized by α. We know that a unique value of Hölder exponent indicates monofractality, while multifractality is identified by different values of α, leading to the existence of the spectrum $f(\alpha)$. Figures 5.5a, 5.5b, 5.6a and 5.6b provide evidence of multifractal

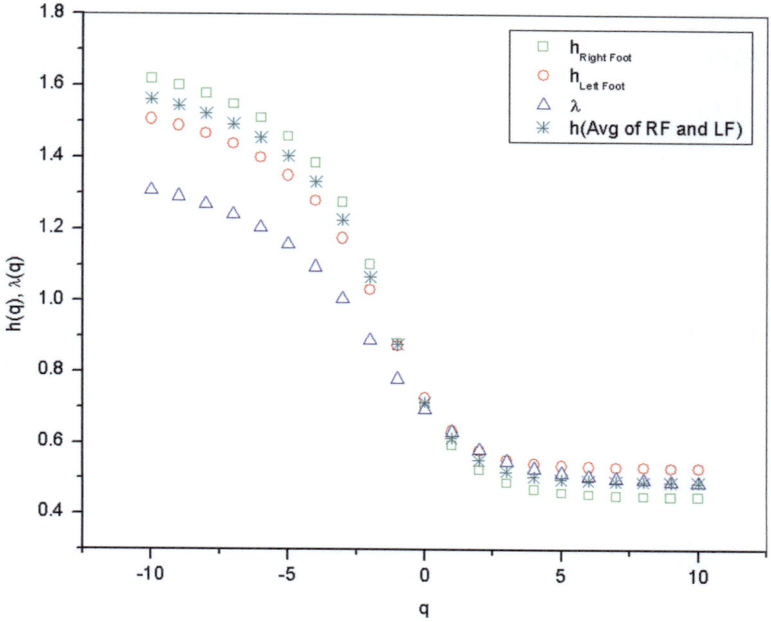

Fig. 5.3a Plot of $h(q)$ and $\lambda(q)$ vs. q for a particular subject of Control Group for Experiment 1 (Dutta et al. 2016)

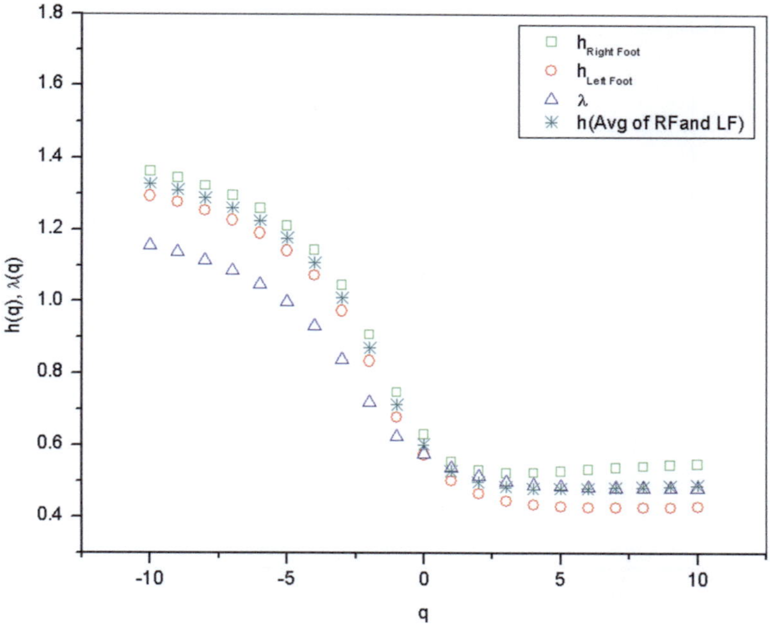

Fig. 5.3b Plot of $h(q)$ and $\lambda(q)$ vs. q for a particular subject of Parkinson's Group for Experiment 1 (Dutta et al. 2016)

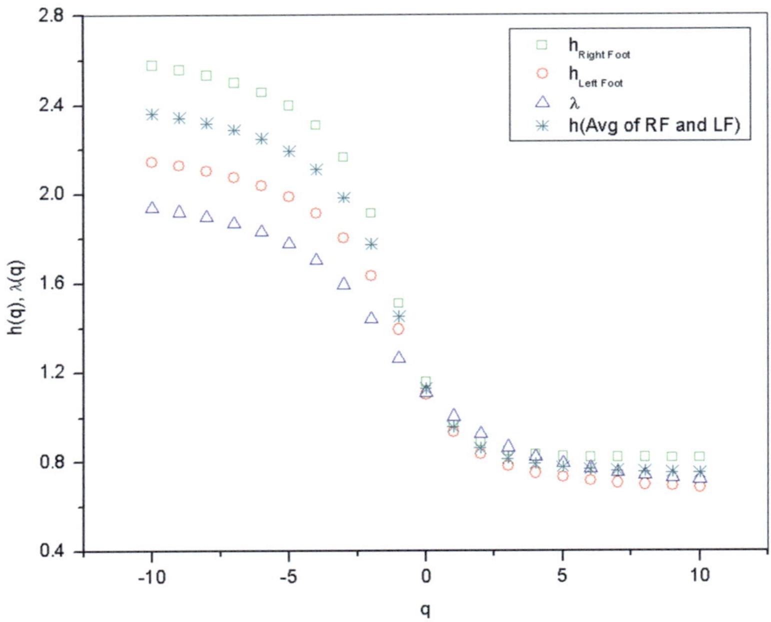

Fig. 5.4a Plot of $h(q)$ and $\lambda(q)$ vs. q for a particular subject of Control Group for Experiment 2 (Dutta et al. 2016)

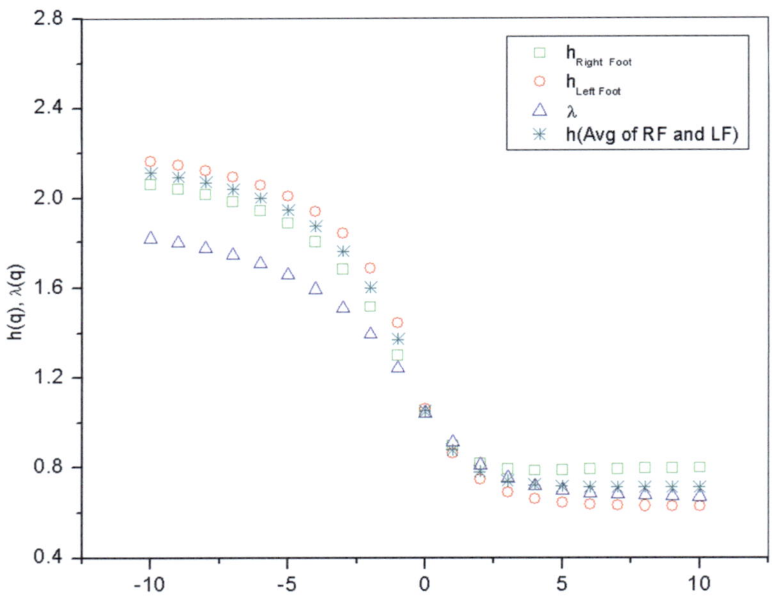

Fig. 5.4b Plot of $h(q)$ and $\lambda(q)$ vs. q for a particular subject of Parkinson's Group for Experiment 2 (Dutta et al. 2016)

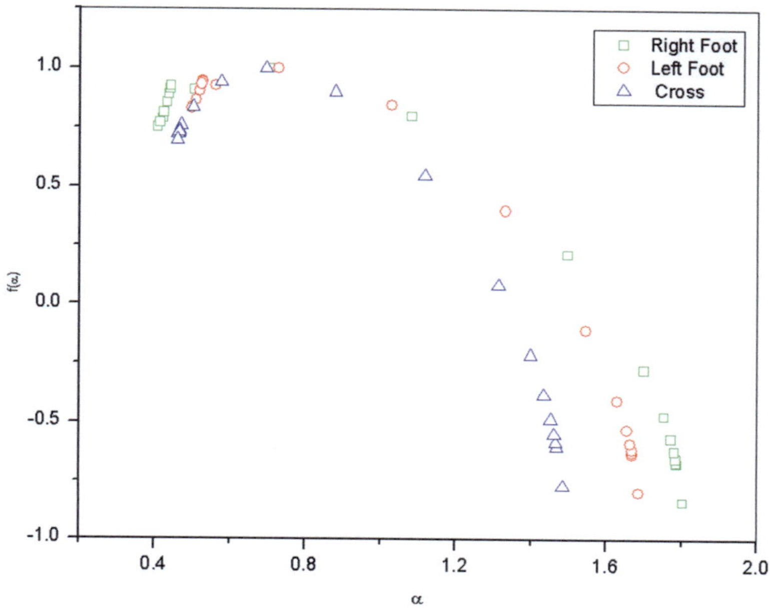

Fig. 5.5a Plot of $f(\alpha)$ vs. α *for a* particular subject of Control Group for Experiment 1 (Dutta et al. 2016)

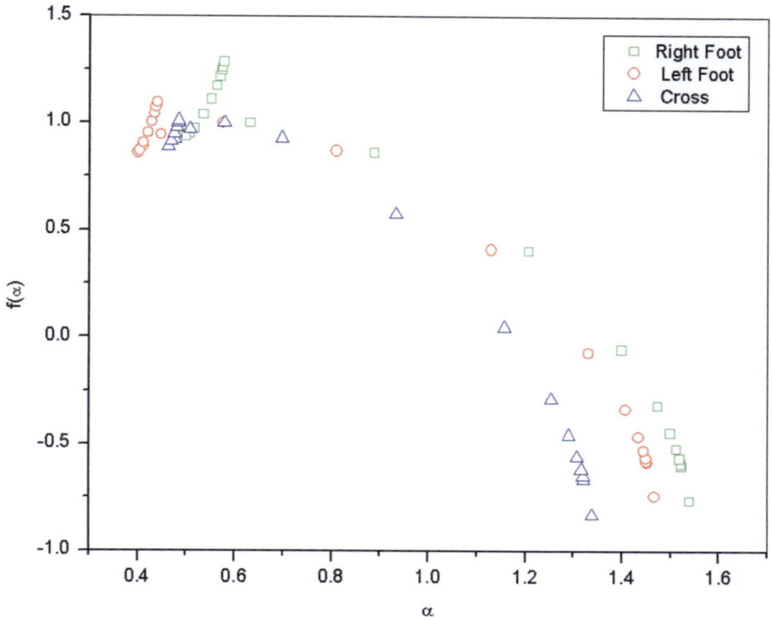

Fig. 5.5b Plot of $f(\alpha)$ vs. α *for a* particular subject of Parkinson's Group for Experiment 1 (Dutta et al. 2016)

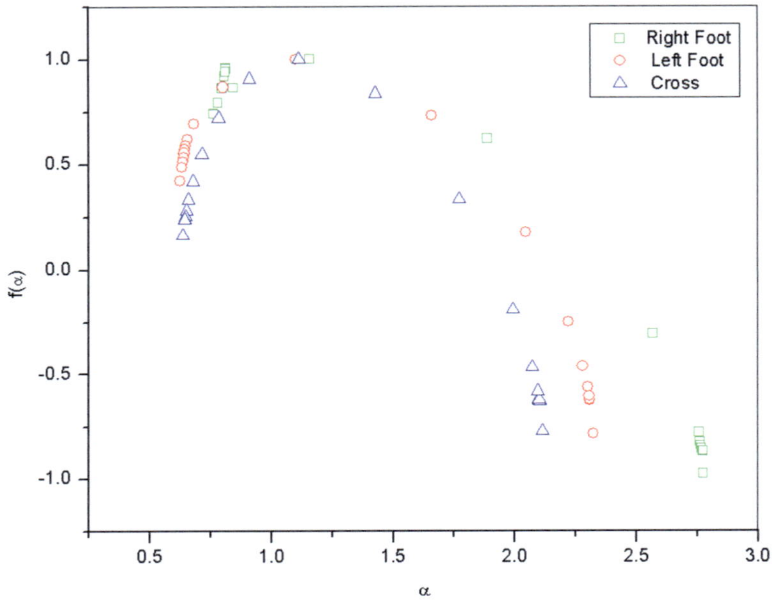

Fig. 5.6a Plot of $f(\alpha)$ vs. α *for a* particular subject of Control Group for Experiment 2 (Dutta et al. 2016)

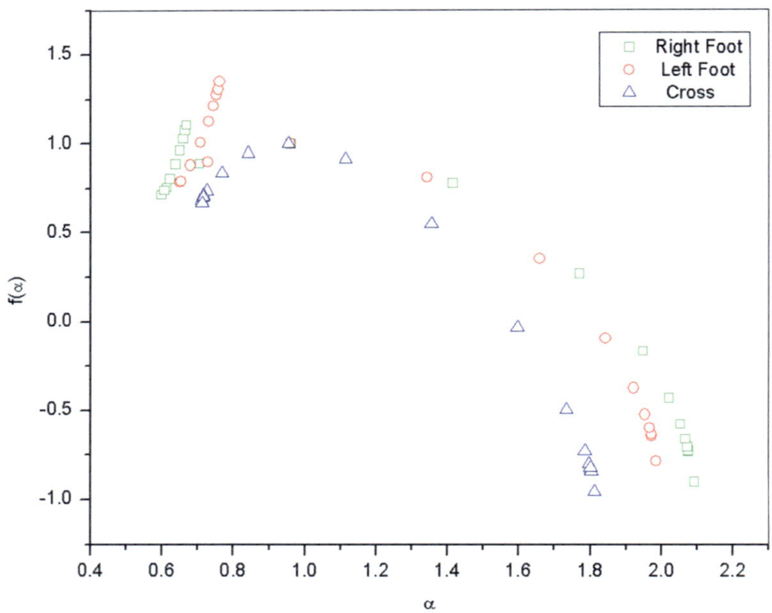

Fig. 5.6b Plot of $f(\alpha)$ vs. α *for a* particular subject of Parkinson's Group for Experiment 2 (Dutta et al. 2016)

behavior for Experiments 1 and 2, respectively. The broadening of the width of the spectrum denotes the increase in degree of multifractality of the signals. For both the experiments, it can be seen that the width of the cross-correlated signal is weaker than the auto-correlated series (Dutta et al. 2016). This phenomenon has previously been observed in other studies too (Movahed and Hermanis 2008; He and Chen 2011; Zhao et al. 2011; Wang and Xie 2012).

The relation where the cross-correlation scaling exponent $\lambda(q)$ is average of the generalized Hurst exponent $[\lambda(q = 2) \approx \{h_x(q = 2) + h_y(q = 2)\}/2]$ of the two time series of left and right foot is found to be between approximately valid in all the cases within the limit of experimental error. The relation is depicted in Figs. 5.3a, 5.3b, 5.4a and 5.4b. The above relation is valid for any value of q for artificially generated time series. However the relation is not valid for any value of q especially for the negative values of q. Oswiecimka et al. (2014) have shown that the more the difference between $\lambda(q)$ and average generalized Hurst exponent, the more different are the considered multifractals. It is evident from Figs. 5.3a, 5.3b, 5.4a and 5.4b that for both the experiments the deviation between $\lambda(q)$ and average generalized Hurst exponent is more for the control group compared to the Parkinson's group (Dutta et al. 2016). We further notice from Figs. 5.3a, 5.3b, 5.4a and 5.4b that for both the experiments the cross-correlation scaling exponent $\lambda(q)$ for $q = 2$ for both healthy persons (Control group) and persons with PD is greater than 0.5 which means that long-range cross-correlation and persistent properties exist in all the sets. A value of $\lambda(q) = 0.5$ signifies nonexistence of cross-correlation, and $\lambda(q) > 0.5$ denotes sustained long-range cross-correlations, i.e., a large value in one variable will be subsequent to a large value in another variable. When $\lambda(q) < 0.5$, anti-persistent cross-correlations are present where a large value in one variable will succeed a small value in another variable and vice versa (Yogev et al. 2005; Movahed and Hermanis 2008). The auto- and the cross-correlation coefficients were estimated for both the experiments. The distribution of values of auto- and cross-correlation coefficients for both control group and diseased set is shown in Figs. 5.7a, 5.7b, 5.8a and 5.8b for Experiments 1 and 2, respectively. The mean values of the auto-correlation coefficient (γ) for the left foot and the right foot and cross-correlation coefficients (γ_x) are listed in Tables 5.1 and 5.2 for Experiments 1 and 2, respectively. The corresponding variances are also listed in the Tables (Dutta et al. 2016).

To determine the statistical significance of the results, we applied ANOVA. The values of F and p and the confidence level of the results being statistically significant are also listed in the tables. A lower value of γ_x is an indication of a higher degree of cross-correlation. It is observed that the right foot and the left foot show almost identical values of auto-correlation in all the cases. The results show that for both the experiments the control group shows a higher degree of auto- and cross-correlation which is in accordance with previous studies of Scafetta et al. (2007) who have observed that due to neuronal deterioration, a network of neurons controlling human gait is less correlated in diseased set than a healthy neuronal network. A leftward shift of the Hölder exponent distribution was observed and was estimated to increase with the severity of the neurodegenerative disease. Hausdorff et al. (1996, 1997, 2000) have also observed loss of correlation in inter-stride interval fluctuation with

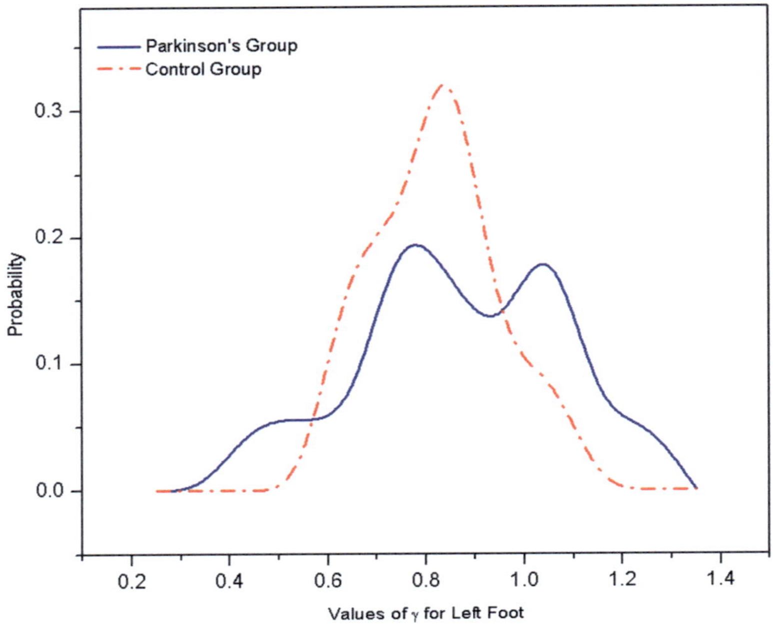

Fig. 5.7a Distribution of values of auto-correlation coefficients for Left foot for Experiment 1 (Dutta et al. 2016)

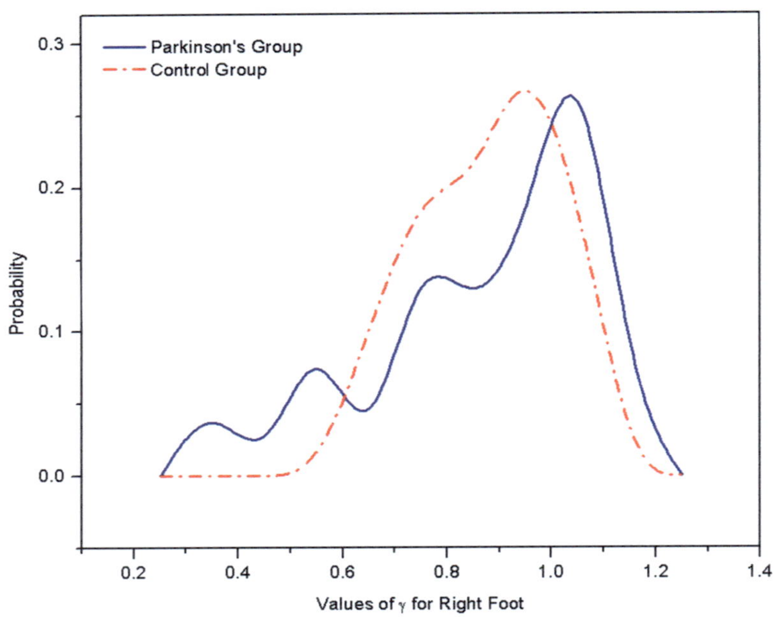

Fig. 5.7b Distribution of values of auto-correlation coefficients for Right foot for Experiment 1 (Dutta et al. 2016)

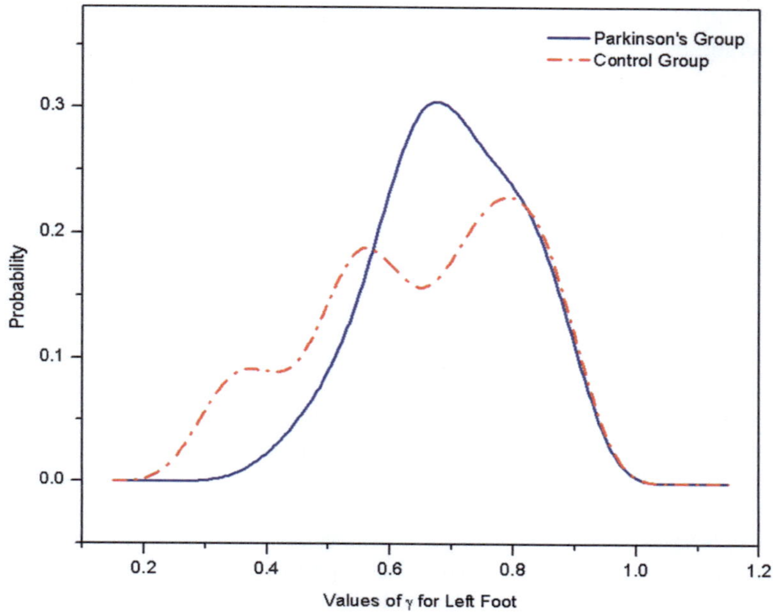

Fig. 5.8a Distribution of values of auto-correlation coefficients for Left foot for Experiment 2 (Dutta et al. 2016)

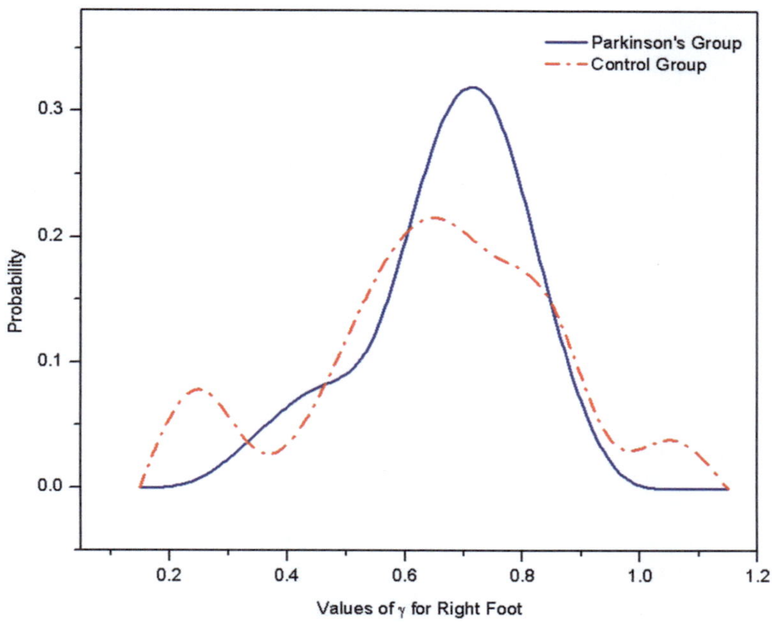

Fig. 5.8b Distribution of values of auto-correlation coefficients for Right foot for Experiment 2 (Dutta et al. 2016)

patients suffering from Parkinson's and Huntington's disease (Dutta et al. 2016). Considering the statistical significance of results, for both groups (control and PD), the cross-correlations show a significant difference in values with a confidence level of about 76% for Experiment 1. But for Experiment 2 the auto-correlation coefficient for the left foot shows most significant distinction (73% confidence level) between the normal and diseased set. Thus it is evident that the auto-correlation coefficients alone are not sufficient to distinguish between normal and diseased set. Cross-correlation coefficients can suggest a discriminate between normal and diseased set when auto-correlation fails to do so (Dutta et al. 2016).

We know that the width of the multifractal spectrum indicates range of exponents. To obtain the values of the exponents, the singularity spectrum is fitted to a quadratic function around the position of maximum α_0. The exponent B measures asymmetry of the spectra. For a symmetric spectrum $B = 0$. A right-skewed spectrum with $B > 0$ indicates dominance of high fractal exponents and hence presence of fine structure, while $B < 0$ suggests smooth structure. From the graphs of multifractal spectrum, we can observe a right-skewed nature for both the experiments for both control and diseased group in most of the cases. In Tables 5.1 and 5.2, the mean values of the parameter B are provided. In Experiment 1 the values of B are significantly different for the cross-correlated series, and control group shows a higher value of B suggesting a more complex structure. However in Experiment 2 no significant difference in value of B is found for control and diseased set (Dutta et al. 2016).

Table 5.1 Mean values of the multifractal width *W* and asymmetry parameter B, auto-correlation coefficients for the left foot and right foot and cross-correlation coefficients for control and Parkinson's group along with variance and ANOVA parameters *F* and *P*, and confidence level for Experiment 1 (Dutta et al. 2016)

Parameter	Group	Mean	Variance	*F*	*P*	Confidence level
γ_L	Control	0.83	0.01	0.32	0.58	42%
	Parkinson's	0.87	0.05			
γ_R	Control	0.86	0.02	0.06	0.82	18%
	Parkinson's	0.88	0.04			
γ_X	Control	0.74	0.02	1.46	0.24	76%
	Parkinson's	0.83	0.04			
W_L	Control	1.56	0.13	0.17	0.68	32%
	Parkinson's	1.62	0.13			
W_R	Control	1.9	0.6	3.72	0.06	94%
	Parkinson's	1.54	0.06			
W_X	Control	1.17	0.04	0.69	0.41	60%
	Parkinson's	1.3	0.2			
B_L	Control	0.21	0.03	0.58	0.45	55%
	Parkinson's	0.27	0.05			
B_R	Control	0.36	0.21	0.18	0.67	33%
	Parkinson's	0.31	0.03			
B_X	Control	0.54	0.03	2.78	0.11	89%
	Parkinson's	0.40	0.05			

Table 5.2 Mean values of the multifractal width *W* and asymmetry parameter B, auto-correlation coefficients for the left foot and right foot and cross-correlation coefficients for control and Parkinson's group along with variance and ANOVA parameters F and P, and confidence level for Experiment 2 (Dutta et al. 2016)

Parameter	Group	Mean	Variance	F	P	Confidence level
γ_L	Control	0.65	0.03	1.24	0.27	73%
	Parkinson's	0.70	0.01			
γ_R	Control	0.64	0.04	0.28	0.60	40%
	Parkinson's	0.67	0.01			
γ_X	Control	0.59	0.03	0.30	0.58	42%
	Parkinson's	0.61	0.01			
W_L	Control	2.4	0.4	1.49	0.23	77%
	Parkinson's	2.1	0.5			
W_R	Control	2.3	0.7	2.07	0.15	85%
	Parkinson's	2.0	0.3			
W_X	Control	1.9	0.2	4.14	0.05	95%
	Parkinson's	1.6	0.1			
B_L	Control	0.26	0.06	0.55	0.46	54%
	Parkinson's	0.33	0.12			
B_R	Control	0.23	0.05	0.62	0.43	57%
	Parkinson's	0.30	0.10			
B_X	Control	0.47	0.05	0.11	0.74	26%
	Parkinson's	0.50	0.09			

In Tables 5.1 and 5.2, the values of degree of multifractality (multifractal width *W*) of auto- and cross-correlated series are presented for Experiments 1 and 2, respectively. These values can also play an important role in differentiating normal and diseased set. Except the control set for Experiment 1, all other sets show almost the same value of degree of multifractality for the right and left foot. For Experiment 1, the degree of multifractality for the right foot shows most significant difference between the normal and diseased set with a confidence level as high as 94%. But for Experiment 2, significant different degrees of multifractality both the auto- and cross-correlated series are observed for normal and diseased set. In most of the cases, the degree of multifractality for control group is more than the patients suffering from PD (Dutta et al. 2016).

Thus the study of cross-correlations between the left and right foot of control subjects and patients suffering from Parkinson's disease (PD) reveals interesting observations. For both healthy and diseased persons, the human gait rhythm is found to exhibit multifractal properties. For all persons with neurodegenerative diseases, the degree of multifractality is significantly less in general. Degree of cross-correlation is stronger in healthy subjects than those suffering from neurodegenerative disease. The study reveals almost same results for auto-correlation and degree of multifractality for the right foot and left foot in most of the cases. Fundamental results obtained from independent Experiments 1 and 2 are almost the same. This leads us to conclude that the parameters W and γ can be used as an index for assessment of onset, severity, and prognosis of different patient suffering from neurodegenerative disease.

Ashkenazy et al. (2002) using wavelet transform modulus maxima and magnitude sign analysis have shown that for healthy adults, stride interval time series is a monofractal. Ivanov et al. (2009) also confirm the same results. The possible reason for us observing a multifractal behavior for healthy adults may arise due to different types of data (ground reaction force) which was used for the study. As discussed earlier Scafetta et al. (2009) expressed complexity of human stride interval which can be described by particular symmetries along with fractal and multifractal properties using the SCPG technique. In elderly or cases with neurodegenerative diseases, they also reported the randomness of fluctuations to be higher (Dutta et al. 2016).

We have used a high range of q from -10 to $+10$ which may trigger spurious multifractality, a limitation discussed by Ivanov et al. (2001). However the results of the present investigation are quite stable with respect to the order of moments. Figure 5.9a and 5.9b is presented to show that the results of multifractal width do not vary much with the order of moments, i.e., when q is varied between -10 to $+10$ and -5 to $+5$.

It has been shown by Podobnik et al. (2007, 2009) that if the auto-correlations are stronger, then the cross-correlations will also be stronger. A χ^2 test would be

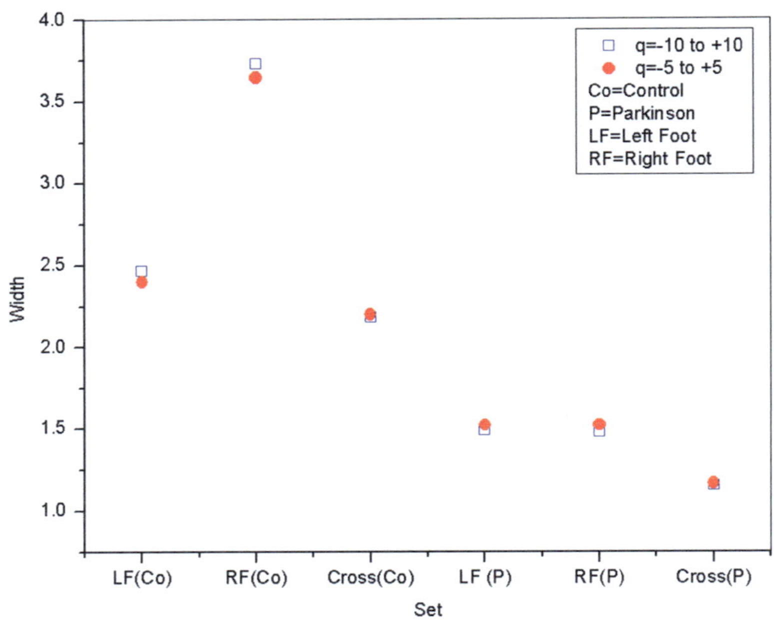

Fig. 5.9a Values of multifractal width W for $q = -10$ to $+10$ and $q = -5$ to $+5$ for Experiment 1 (Dutta et al. 2016)

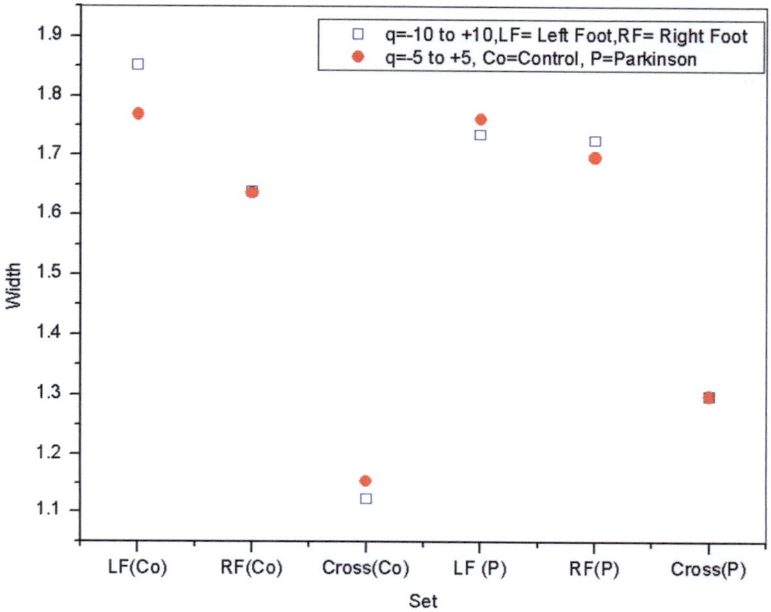

Fig. 5.9b Values of multifractal width W for $q = -10$ to +10 and $q = -5$ to +5 for Experiment 2 (Dutta et al. 2016)

indicative of presence of strong cross-correlations in the series. The results are sensitive to order of detrending (Oswiecimka et al. 2013). A comparison of results of different order of detrending would give more appropriate results. Nevertheless the study presents new data on the cross-correlations of human gait time series. More analysis with different data sets of various neurodegenerative diseases will be helpful not only for better understanding of the dynamics of neurodegenerative disease but also can be applied as a medical diagnostic tool (Dutta et al. 2016).

In the next section, we will focus on gait dynamics of another neurodegenerative disease, namely, Huntington's disease (HD).

5.4 Huntington's Disease and Gait Data

Huntington's disease (HD) is a neurodegenerative genetic disorder that affects muscle coordination and leads to cognitive decline and psychiatric problems. The disease is progressive in nature which causes physical, mental, and emotional alterations. The people affected cannot think, talk, and move properly. Basal ganglia cells get damaged, which controls these capacities. HD being hereditary, there is 50% possibility that a child may acquire the abnormal gene. In the initial phase of the disease, signs of poor memory; problems to take decisions; mood changes like depression, anger, or irritability; increasing lack of coordination, twitching, or other uncontrolled movements; and difficulty in walking, speaking, or swallowing are seen (Singh et al. 2013). Grimbergen et al. (2008) examined fall and gait disturbances in Huntington's disease. On comparing with healthy control, decreased stride length and decreased gait velocity were observed in Huntington's patients. Since fallers were found to possess both the characteristics, falling was assumed to be the common syndrome of Huntington's disease. Huntington's disease cannot heal completely. With medicine some symptoms can only be controlled, not eradicate the disease.

In order to investigate the fractal properties of the human gait in case of normal persons (control group) and patients with Parkinson's and Huntington's diseases, we studied the databases of human gait from the website www.physionet.org. The records in the neurodegenerative disease are from patients with Parkinson's disease and Huntington's disease, and records from healthy subjects (Control Group) have been included as the comparison group (Dutta et al. 2013)

5.5 Multifractal Analysis of Huntington's Data

An interesting work was reported by Dutta, Ghosh, and Chatterjee (Dutta et al. 2013) where a multifractal analysis of gait time series of two diseased set, namely, Parkinson's and Huntington's, and a control group was reported using MF-DFA (the details of the method are discussed in Appendix A). We are aware that normal gait time series is highly inhomogeneous and nonstationary and fluctuates about the mean value in an irregular and complex manner.

The three sets of data were first transformed into the integrated signal and then divided into N_s nonoverlapping bins. The value of s was chosen in the range 5 to $N/5$ in steps of 1 ($N = 1200$) where s is the length of the bin and N length of the time series. By varying values of q from -10 to +10, the fluctuation function $F_q(s)$ was obtained. For different q values, linear dependence of fluctuation function on scale s for the three groups, namely, (i) control group, (ii) persons with Parkinson's disease, and (iii) persons with Huntington's disease, was observed indicating a scaling behavior. We can get values of generalized Hurst exponent $h(q)$ from the plot of linear fit of fluctuation function and time scale s. The obtained Hurst exponent $h(q)$ values were found to vary with q indicating a multifractal behavior. On calculating the values of classical scaling exponent $\tau(q)$, it was found that $\tau(q)$ depends non-linearly on q which gives evidence of multifractality of all the considered time series. Thus both the variation of Hurst exponent and classical scaling exponent with q reflect multifractality in human gait for each of healthy group and patients suffering from Parkinson's and Huntington's disease. For each of the data set, the degree of multifractality was determined quantitatively by calculating values of singularity strength α and spectrum $f(\alpha)$ the dimension of subset series. The width W of the spectrum is a measure of the degree of multifractality. The mean values of width W of the multifractal spectra listed in Table 5.3 show that the width of

Table 5.3 Mean values, variance of multifractal width *W*, and ANOVA parameters F and p values for all three groups (Dutta et al. 2013)

Set	Values of *W*		*F*	*P*
	Average	Variance		
Control left foot	3.7	2.1	8.79	0.002
Parkinson's left foot	2.2	0.1		
Huntington's left foot	2.15	0.03		
Control right foot	3.8	1.2	6.25	0.008
Parkinson's right foot	2.8	1.1		
Huntington's right foot	2.3	0.3		

Table 5.4 Values of multifractal width *W* and auto-correlation coefficient γ for (i) healthy subjects (control group), (ii) subjects with Parkinson's disease, and (iii) subjects persons with Huntington's disease for both the original series and the shuffled series (Dutta et al. 2013)

SET	FOOT	W_{original}	W_{shuffled}	γ_{original}	γ_{shuffled}
Control group	Left	4.0 ± 0.1	0.93 ± 0.07	0.53	1.03
	Right	4.0 ± 0.2	1.2 ± 0.2	0.43	1.06
Parkinson's disease	Left	2.3 ± 0.2	0.88 ± 0.07	0.67	0.99
	Right	2.7 ± 0.1	0.93 ± 0.07	0.61	0.99
Huntington's disease	Left	2.20 ± 0.09	0.90 ± 0.08	0.82	1.03
	Right	2.07 ± 0.09	0.72 ± 0.05	0.86	1.07

multifractal spectrum is greater in case of the healthy subject (control group) than those for patients with Parkinson's and Huntington's disease which suggests that the degree of multifractality is more in case of healthy subjects than those with neurodegenerative diseases (Dutta et al. 2013).

To ascertain the origin of multifractality, the corresponding randomly shuffled series was analyzed following the same procedure as for the original series for all the three cases. In Table 5.4 the values of $W_{shuffled}$ and $\gamma_{shuffled}$ (auto-correlation shuffled), the width of the multifractal spectra, and correlation coefficient for the original and shuffled series for one subject in each group are depicted. Comparing the values one can suggest that the origin of multifractality is due to both broad probability distribution and long-range correlation, though long-range correlation is dominant as evident from reduced values of multifractal width. Compared to auto-correlation (γ) values for the original series, the values of $\gamma_{shuffled}$ are close to 1 as expected since all correlations are destroyed in the shuffling procedure. The values of the auto-correlation coefficient reveal that the gait series is long-range positive correlated series which approaches toward an uncorrelated series with neurodegenerative diseases (Dutta et al. 2013).

Thus the analysis of human gait rhythm with MF-DFA technique of healthy persons and persons with Parkinson's diseases and Huntington disease reveals that the gait rhythm manifests multifractal properties. The multifractal properties are found to be more pronounced in normal persons, i.e., degree of multifractality is greater in normal persons, than in persons with neurodegenerative diseases. Thus the MF-DFA method is capable of distinguishing between normal and diseased set. In almost all of the cases, the left foot and the right foot data are seen to produce identical results (Dutta et al. 2013).

Using ANOVA statistical significance of the results was evaluated. The values of F and p are listed in Table 5.3. The values of *W* are found to be different in normal and diseased set with a confidence level about 95%. Thus we can infer that the neurodegenerative diseases can bring about an alteration in the fractal dynamics of human gait due to weakening and impairment of neural control on locomotion (Dutta et al. 2013). The results are consistent with previous studies. Hausdorff et al. (1996, 1997, 2000, 2001, Goldberger et al. 2002b) have observed loss of correlation in inter-stride interval fluctuation with patients suffering from PD and HD. Scafetta et al. (2007) have observed that due to neuronal deterioration, a network of neurons controlling human gait is expected to be less correlated in diseased set than a healthy neuronal network.

5.6 Discussions on Possible Use of the Result for Biomarkers of Parkinson's and Huntington's

Neurodegenerative disease like Parkinson's disease and Huntington's disease can not only be analyzed by studying EEG signals, but gait analysis can also provide useful information in detection of these diseases. From literature it is very clear that subjects with neurodegenerative diseases encounter gait problem. Cessation of movement, shuffling of steps, taking small steps, etc. are some of the acute symptoms of these diseases. Hence researchers have not only focused on EEG related study but also have researched gait dynamics with effective results. Since manual estimation of stride interval is a very lengthy process and is also prone to errors, thus it is the need of the day to develop automated techniques to detect heel and toe strikes. Computational technique can measure heel and toe strike interval. Heel and toe strike time interval of neurodegenerative disease subjects is then to be compared with healthy subjects to get an estimate of the severity of these diseases. Study of these diseases in the light of non-linear dynamics can provide a proper estimate of degree of auto-correlation and cross-correction in the gait rhythm which can act as possible biomarker for detection of Parkinson's and Huntington's disease.

Acknowledgment The authors gratefully acknowledge Physica A and Elsevier Publishing Co. for providing the copyrights of Figs. 5.2a, 5.2b, 5.2c, 5.2d, 5.2e and 5.2f, 5.3a, 5.3b, 5.4a, 5.4b, 5.5a, 5.5b, 5.6a, 5.6b, 5.7a, 5.7b, 5.9a, 5.9b and Tables 5.1 and 5.2 for use in this chapter.

References

Ashkenazy Y, Hausdorff JM, Ivanov P, Goldberger AL, Stanley HE (2002) A stochastic model of human gait dynamics. Phys A 316:662–670

Baltadjieva R, Giladi N, Gruendlinger L, Peretz C, Hausdorff JM (2006) Marked alterations in the gait timing and rhythmicity of patients with de novo Parkinson's disease. Eur J Neurosci 24:1815–1820

Beauchet O, Herrmann FR, Grandjean R, Dubost V, Allali G (2008) Concurrent validity of SMTEC® footswitches system for the measurement of temporal gait parameters. Gait Posture 27:156–159

Blin O, Ferrandez AM, Serratrice G (1990) Quantitative analysis of gait in Parkinson patients: increased variability of stride length. J Neurol Sci 98:91–97

Bunde A, Knopp J, Schellnhuber HJ (eds) (2002) The science of disasters. Springer, Berlin

Buzzi UH, Ulrich BD (2004) Dynamic stability of gait cycles as a function of speed and system constraints. Mot Control 8:241–254

Cavanaugh JT, Coleman KL, Gaines JM et al (2007) Using step activity monitoring to characterize ambulatory activity in community-dwelling older adults. J Am Geriatr Soc 55:120–124

Collins JJ, Richmond SA (1994) Hard-wired central pattern generators for quadrupedal locomotion. Biol Cybern 71:375–385

Collins JJ, Stewart IN (1993) Coupled non-linear oscillators and the symmetries of animal gaits. J Nonlinear Sci 3:349–392

Costa M, Peng CK, Goldberger AL, Hausdorff JM (2003) Multiscale entropy analysis of human gait dynamics. Phys A 330:53–60

Davis R, Õunpuu S, Tyburski D, Gage JR (1991) A gait analysis data collection and reduction technique. Hum Mov Sci 10:575–587

Decker LM, Cignettia F, Stergiou N (2010) Complexity and human gait. Revista Andaluza de Medicina del Deporte 3:2–12

Delignières D, Torre K (2009) Fractal dynamics of human gait: a reassessment of the 1996 data of Hausdorff et al. J Appl Physiol 106:1272–1279

Dingwell JB, Cusumano JP (2000) Non-linear time series analysis of normal and pathological human walking. Chaos 10:848–863

Dingwell JB, Kang HG (2007) Differences between local and orbital dynamic stability during human walking. J Biomech Eng 129:586–593

Dutta S, Ghosh D, Chatterjee S (2013) Multifractal Detrended fluctuation analysis of human gait diseases. Front Physiol 4:274

Dutta S, Ghosh D, Samanta S (2016) Non linear approach to study the dynamics of neurodegenerative diseases by multifractal Detrended cross-correlation analysis—a quantitative assessment on gait disease. Phys A 448:181–195

Feder J (1988) Fractals. Plenum Press, New York

Frenkel-Toledo S, Giladi N, Peretz C, Herman T, Gruendlinger L et al (2005a) Treadmill walking as an external pacemaker to improve gait rhythm and stability in Parkinson's disease. Mov Disord 20:1109–1114

Frenkel-Toledo S, Giladi N, Peretz C, Herman T, Gruendlinger L et al (2005b) Effect of gait speed on gait rhythmicity in Parkinson's disease: variability of stride time and swing time respond differently. J Neuroeng Rehabil 2:23

Gabell A, Nayak US (1984) The effect of age on variability in gait. J Gerontol 39:662–666

Goldberger AL, Peng CK, Lipsitz LA (2002a) What is physiologic complexity and how does it change with aging and disease? Neurobiol Aging 23:23–26

Goldberger AL, Amaral LA, Hausdorff JM, Ivanov PC, Peng CK et al (2002b) Fractal dynamics in physiology: alterations with disease and aging. Proc Natl Acad Sci U S A 99:2466–2472

Griffin L, West DJ, West BJ (2000) Random stride intervals with memory. J Biol Phys 26:185

Grimbergen YA, Knol MJ, Bloem BR, Kremer BP, Roos RA et al (2008) Falls and gait disturbances in Huntington's disease. Mov Disord 23:970–976

Guimares RM, Isaacs B (1980) Characteristics of the gait in old people who fall. Int Rehabil Med 2:177–180
Harbourne RT, Stergiou N (2009) Movement variability and the use of non-linear tools: principles to guide physical therapy practice. Phys Ther 89:267–282
Hausdorf JM, Alexander NB (2005) Gait disorders: evaluation and management. Informa Healthcare, New York
Hausdorf JM, Mitchell SL, Firtion R, Peng CK, Cudkowiczet ME, al. (1997) Altered fractal dynamics of gait: reduced stride-interval correlations with aging and Huntington's disease. J Appl Physiol 82:262–269
Hausdorf JM, Ashkenazy Y, Peng CK, Ivanov PC, Stanley HE et al (2001) When human walking becomes random walking: fractal analysis and modeling of gait rhythm fuctuations. Phys A 302:138–147 and references cited therein
Hausdorf JM, Lowenthal J, Herman T, Gruendlinger L, Peretz C et al (2007) Rhythmic auditory stimulation modulates gait variability in Parkinson's disease. Eur J Neurosci 26:2369–2375
Hausdorff JM (2007) Gait dynamics, fractals and falls: finding meaning in the stride-to-stride fluctuations of human walking. Hum Mov Sci 26:555–589
Hausdorff JM, Peng CK, Ladin Z, Wei JY, Goldberger AL (1995) Is walking a random walk? Evidence for long-range correlations in stride interval of human gait. J Appl Physiol 78:349–358
Hausdorff JM, Purdon PL, Peng CK, Ladin Z, Wei JY et al (1996) Fractal dynamics of human gait: stability of long-range correlations in stride interval fluctuations. J Appl Physiol 80:1448–1457
Hausdorff JM, Cudkowicz ME, Firtion R, Wei JY, Goldberger AL (1998) Gait variability and basal ganglia disorders: stride-to-stride variations of gait cycle timing in Parkinson's disease and Huntington's disease. Mov Disord 13:428–437
Hausdorff JM, Zemany L, Peng CK, Goldberger AL (1999) Maturation of gait dynamics: stride-to-stride variability and its temporal organization in children. J Appl Physiol 86:1040–1047
Hausdorff JM, Lertratanakul A, Cudkowicz ME, Peterson AL, Kaliton D, Goldberger AL (2000) Dynamic markers of altered gait rhythm in amyotrophic lateral sclerosis. J Appl Physiol 88:2045–2053
He LY, Chen SP (2011) Multifractal Detrended cross-correlation analysis of agricultural futures markets. Chaos, Solitons Fractals 44:355–361
Herman T, Giladi N, Gurevich T, Hausdorff JM (2005) Gait instability and fractal dynamics of older adults with a cautious gait: why do certain older adults walk fearfully? Gait Posture 21:178–185
Hillmana SJ, Stansfieldb BW, Richardsonc AM, Robb JE (2009) Development of temporal and distance parameters of gait in normal children. Gait Posture 29:81–85
Holt KG, Saltzman E, Ho CL, Ulrich BD (2007) Scaling of dynamics in the earliest stages of walking. Phys Ther 87:1458–1467
Hurst HE (1951) Long-term storage capacity of reservoirs. Trans Am Soc Civ Eng 116:770–808
Inman VT, Ralston HJ, Todd F (1981) Human Walking. Williams & Wilkins, Baltimore, p 154
Inman VT, Ralston HJ, Todd F (2006) In: Rose J, Gamble Lippincott JG (eds) Human walking. Williams & Wilkins, Philadelphia, pp 7–18
Ivanov PC, Amaral LAN, Goldberger AL, Havlin S, Rosenblum MG et al (1999) Multifractality in human heartbeat dynamics. Nature 399:461–465
Ivanov P, Amaral LAN, Goldberger AL, Havlin S, Rosenblum MG et al (2001) From 1/f noise to multifractal cascades in heartbeat dynamics. Chaos 11:641–652
Ivanov PC, Ma QDY, Bartsch R, Hausdorff JM, Amaral LAN et al (2009) Levels of complexity in scale-invariant neural signals. Phys Rev E 79:041920
Jordan K, Challis JH, Newell KM (2006) Long range correlations in the stride interval of running. Gait Posture 24:120–125
Jordan K, Challis JH, Newell KM (2007a) Walking speed influences on gait cycle variability. Gait Posture 26:128–134
Jordan K, Challis JH, Newell KM (2007b) Speed influences on the scaling behavior of gait cycle fluctuations during treadmill running. Hum Mov Sci 26:87–102

Kadaba MP, Ramakrishnan HK, Wootten ME, Gainey J, Gorton G et al (1989) Repeatability of kinematic, kinetic, and electromyographic data in normal adult gait. J Orthop Res 7:849–860
Khandoker AH, Palaniswami R, Begg RK (2008) A comparative study on approximate entropy measure and poincaré plot indexes of minimum foot clearance variability in the elderly during walking. J Neuroeng Rehabil 5:4
Kuo AD (2002) The relative roles of feedforward and feedback in the control of rhythmic movements. Mot Control 6:129
Kurz M, Stergiou N (2003) The aging human neuromuscular system expresses less certainty for selecting joint kinematics during gait. Neurosci Lett 348:155–158
Kurz M, Stergiou N (2007) Do horizontal propulsive forces influence the non-linear structure of locomotion? J Neuroeng Rehabil 1:4–30
Lomax RG (2007) Statistical concepts: a second course for education and the behavioral sciences. Lawrence Erlbaum Associates, Mahwah
Malamud BD, Turcotte DL (1999) Self-affine time series: i. generation and analysis. Adv Geophys 40:1–90
Matjaz P (2005) The dynamics of human gait. Eur J Phys 26:525
Miller RA, Thaut MH, McIntosh GC, Rice RR (1996) Components of EMG symmetry and variability in parkinsonian and healthy elderly gait. Electroencephalogr Clin Neurophysiol 101:1–7
Miller DJ, Stergiou N, Kurz MJ (2006) An improved surrogate method for detecting the presence of chaos in gait. J Biomech 39:2873–2876
Moraiti C, Stergiou N, Ristanis S, Georgoulis AD (2007) ACL deficiency affects stride-to-stride variability as measured using non-linear methodology. Knee Surg Sports Traumatol Arthrosc 15:1406–1413
Morris ME, Iansek R, Matyas TA, Summers JJ (1994a) The pathogenesis of gait hypokinesia in Parkinson's disease. Brain 117:1169–1181
Morris ME, Iansek R, Matyas TA, Summers JJ (1994b) Ability to modulate walking cadence remains intact in Parkinson's disease. J Neurol Neurosurg Psychiatry 57:1532–1534
Morris ME, Iansek R, Matyas TA, Summers JJ (1996a) Stride length regulation in Parkinson's disease. Normalization strategies and underlying mechanisms. Brain 119:551–568
Morris ME, Matyas TA, Iansek R, Summers JJ (1996b) Temporal stability of gait in Parkinson's disease. Phys Ther 76:763–777
Morris ME, McGinley J, Huxham F, Collier J, Iansek R (1999) Constraints on the kinetic and spatiotemporal parameters of gait in Parkinson's disease. Hum Mov Sci 18:461–483
Morris ME, Huxham F, McGinley J, Dodd K, Iansek R (2001) The biomechanics and motor control of gait in Parkinson disease. Clin Biomech 16:459–470
Movahed S, Hermanis E (2008) Fractal analysis of river flow fluctuations. Phys A 387:915–932
Muñoz-Diosdado A (2005) A non linear analysis of human gait time series based on multifractal analysis and cross correlations. J Phys Conf Ser 23:87–95
Oswiecimka P, Drożdż S, Kwapień J, Górski A (2013) Effect of detrending on multifractal characteristics. Acta Phys Pol A 123:597–603
Oswiecimka P, Drożdż S, Forczek M, Jadach S, Kwapien J et al (2014) Detrended cross-correlation analysis consistently extended to multifractality. Phys Rev E 89:023305
Pailhous J, Bonnard M (1992) Steady-state fluctuations of human walking. Behav Brain Res 47:181–190
Palta AE (1985) Some characteristics of EMG patterns during locomotion: implications for locomotor control processes. J Mot Behav 17:443–461
Peitgen HO, Jurgens H, Saupe D (1992) Chaos and fractals. Springer-Verlag, New York Appendix B
Podobnik B, Fu DF, Stanley HE, Ivanov PC (2007) Power-law autocorrelated stochastic processes with long-range cross-correlations. Eur Phys J B 56:47–52
Podobnik B, Grosse I, Horvatic D, IIic S, Ivanov PC et al (2009) Quantifying cross-correlations using local and global detrending approaches. Eur Phys J B 71:243–250

Robbins SM, Astephen Wilson JL, Rutherford DJ, Hubley-Kozey CL (2013) Reliability of principal components and discrete parameters of knee angle and moment gait waveforms in individuals with moderate knee osteoarthritis. Gait Posture 38:421–427
Scafetta N, Griffin L, West BJ (2003) Hölder exponent spectra for human gait. Phys A 328:561–583
Scafetta N, Moon RE, West BJ (2007) Fractal response of physiological signals to stress conditions, environmental changes, and neurodegenerative diseases. Complexity 12:12–17
Scafetta N, Marchi D, West BJ (2009) Understanding the complexity of human gait dynamics. Chaos 19:026108–026110
Sekine M, Tamura T, Akay M, Fujimoto T, Togawa T et al (2002) Discrimination of walking patterns using wavelet-based fractal analysis. IEEE Trans Neural Syst Rehabil Eng 10:188–196
Sekine M, Akay M, Tamura T, Higashi Y (2004) Fractal dynamics of body motion in patients with Parkinson's disease. J Neural Eng 1:8–15
Singh M, Singh M, Paramjeet (2013) Neuro-degenerative disease diagnosis using human gait: a review. IJITKMI 7:16–20
Smith BA, Teulier C, Sansom J, Stergiou N, Ulrich BD (2011) Approximate entropy values demonstrate impaired neuromotor control of spontaneous leg activity in infants with myelomeningocele. Pediatr Phys Ther 23(3):241
Stergiou N, Decker LM (2011) Human movement variability, non-linear dynamics, and pathology: is there a connection? Hum Mov Sci 30:869–888
Stergiou N, Buzzi UH, Kurz MJ, Heidel J (2004) Non-linear tools in human movement. In: Stergiou N (ed) Innovative analyses of human movement. Human Kinetics, Champaign, pp 63–90
Stolze H, Kuhtz-Buschbeck JP, Drucke H, Johnk K, Illert M et al (2001) Comparative analysis of the gait disorder of normal pressure hydrocephalus and Parkinson's disease. J Neurol Neurosurg Psychiatry 70:289–297
Tochigi Y, Segal NA, Vaseenon T, Brown TD (2012) Entropy analysis of tri-axial leg acceleration signal waveforms for measurement of decrease of physiological variability in human gait. J Orthop Res 30:879–904
Torres BDLC, Sánchez López MD, Sarabia Cachadiña E, Naranjo Orellana J (2013) Entropy in the analysis of gait complexity: a state of the art. Br J Appl Sci Technol 3:1097–1105
Van Emmerik REA, Rosenstein MT, McDermott WJ, Hamill J (2004) Non-linear dynamical approaches to human movement. J Appl Biomech 20:396–420
Van Orden GC, Kloos H, Wallot S (2009) Living in the pink: intentionality, wellbeing, and complexity. In: Handbook of the philosophy of science volume 10: philosophy of complex systems, vol 10. Elsevier, Amsterdam, pp 639–682
Wang GJ, Xie C (2012) Cross-correlations between WTI crude oil market and US stock market: a perspective from Econophysics. Acta Phys Pol B 43:2021–2036
Webster KE, Wittwer JE, Feller JA (2005) Validity of the GAITRite walkway system for the measurement of averaged and individual step parameters of gait. Gait Posture 22:317–321
West BJ, Scafetta N (2003) Non-linear dynamical model of human gait. Phys Rev E 67:051917
Winter DA (1984) Kinematic and kinetic patterns in human gait: variability and compensating effects. Hum Mov Sci 3:51–76
Winters JM, Crago PE (2000) Biomechanics and neural control of posture and movements. Springer, New York
Wu Y, Krishnan S (2010) Statistical analysis of gait rhythm in patients with Parkinson's disease. IEEE Trans Neural Syst Rehabil Eng 18:150–158
Yamasaki M, Sasaki T, Tsuzki S, Torii M (1984) Stereotyped pattern of lower limb movement during level and grade walking on treadmill. Ann Physiol Anthropol 3:291–296
Yamasaki M, Sasaki T, Torii M (1991) Sex difference in the pattern of lower limb movement during treadmill walking. Eur J Appl Physiol Occup Physiol 62:99–103
Yogev G, Giladi N, Peretz C, Springer S, Simon ES et al (2005) Dual tasking, gait rhythmicity, and Parkinson's disease: which aspects of gait are attention demanding? Eur J Neurosci 22:1248–1256

Yoshino K, Motoshige T, Araki T, Matsuoka K (2004) Effect of prolonged free-walking fatigue on gait and physiological rhythm. J Biomech 37:1271–1280
Yu ZG, Wang B (2001) A time series model for CDS sequences in complete genome. Chaos, Solitons Fractals 12:519–526
Zhao X, Shang P, Jin Q (2011) Multifractal detrended cross-correlation analysis of Chinese stock markets based on time delay. Fractals 19:329–338
Zhou WX (2008) Multifractal detrended cross-correlation analysis for two nonstationary signals. Phys Rev E 77:066211

Chapter 6
Multifractal Correlation Study Between Posture and Autonomic Deregulation Using ECG and Blood Pressure Data

Abstract Recently posture-induced cardiovascular changes have become a subject of study for its obvious implications. This chapter presents analysis of non-linear time series of ECG and arterial blood pressure (ABP) data from a new perspective using state-of-the-art non-linear technique for assessment of cardiac disorder induced by different human posture. Precisely two methods multifractal detrended fluctuation analysis and multifractal detrended cross-correlation analysis have been applied to ECG and ABP data to quantify degree of impact of posture on cardiovascular functions. Moreover the knowledge of cross-correlation parameters is significant as it provides information for understanding the dynamics of orthostatic stress.

6.1 Introduction

The cardiovascular system, the heart, and circulation are regulated by higher brain centers. The parts of the brain that modulate the cardiovascular system stem through the activity of sympathetic and parasympathetic nerves (Hainsworth 1998). Cardiovascular variability analysis permits insight into the neural control mechanism of the heart, leading to a new discipline known as "neurocardiology" (Natelson 1985; Malik and Camm 1990; Aubert and Ramaekers 1999). The autonomic nervous system (ANS) belongs to the central and peripheral nervous systems and is not controlled by our wills. Due to the meaningful and noninvasive characteristics of heart rate variability (HRV), it is used to determine the ANS activities of human conventionally (Sornmo and Laguna 2005; Thayer et al. 2010, 2012).

As emphasized by Acharya et al. (2006), to study cardiovascular control physiology and heartbeat dynamics, it is essential to perform mathematical modeling as well as digital signal processing techniques. An estimate of the time interval between two successive R waves recorded from the ECG signal, i.e., the RR interval, is used in studying the cardiovascular system (Valenza et al. 2015). Since the heartbeat is known to be regulated by the autonomous nervous system, the RR interval shows oscillations about the mean value and is known as heart rate variability (HRV) (Acharya et al. 2006).

D. Ghosh et al., *Multifractals and Chronic Diseases of the Central Nervous System*,
https://doi.org/10.1007/978-981-13-3552-5_6

It is interesting to note that besides respiration, normal heartbeat and blood pressure depend on factors like physical, environmental, mental, etc. and are characterized by circadian variation. Alterations in the autonomic activity determine both the heart rate and its modulation (Aubert et al. 2003). The autonomic nervous system (ANS) is comprised of sympathetic nervous system and parasympathetic nervous system. The activation of the sympathetic nervous system can raise blood pressure and accelerate the heart rate and the parasympathetic nervous system can slow down the heart rate (Tseng et al. 2013). Thus the autonomic nervous system is vital in modulating the cardiovascular system by performing maximum function during different physical activities in healthy people and also by demonstrating different diseases of the cardiac system (Aubert et al. 2003).

Various physiological procedures have been used to get a deeper perception of the functioning of the autonomic nervous system. Investigators have often used a change in posture by either passive head-up tilt or active standing to impose perturbation on the steady-state functioning of the ANS (Nepal and Paudel 2012). For a change in body position from lying down to standing up under normal physiological conditions, the body's internal controlling process curbs large changes of cerebral perfusion and cerebral blood flow during the action. Both cerebral perfusion and cerebral blood flow are not maintained sufficiently in patients with abnormal function of cerebral autoregulation and ANS disorder. Change in cerebral blood flow velocity (CBFV), blood pressure (BP), and heartbeat or heart rate (HR) due to change of posture can be studied performing tilt table test (Tseng et al. 2013).

Postural hypotension patients may encounter whirling sensation or faint when they get up suddenly from resting position. In such situation tilt table test can be used which can provide information regarding the possible causes and severities of postural hypotension. A change in blood pressure during tilt up may give an indication of orthostatic hypotension. Additionally confiding on the condition of ANS and cerebral autoregulation, patients' heart rate and cerebral blood flow velocity may or may not change during tilt up. In these situations analysis of patients' HR, BP, and CBFV changes and also interactions between these signals becomes significant. This comprehensive approach can be helpful in evaluating the patients' ANS and cerebral autoregulation (Tseng et al. 2013). Thus body position and postural changes determine a gravitational gradient acting upon the cardiovascular and pulmonary systems (Jones et al. 2003). The maintenance of upright posture not only requires coordinated neuromuscular control of postural muscles (Winter et al. 1996) but also cardiovascular reflexes to maintain BP. The analysis of variability of these biological signals using linear statistics (mean values, variability measures, and spectra analysis) does not directly characterize their complexity, irregularity, or predictability. Methods based on non-linear dynamics and "chaos" theories may reveal subtle abnormalities in the cardiovascular regulation mechanisms that may not be uncovered by traditional linear measures of variability, and thus, they provide a useful tool for in-depth evaluation of the properties of complex biological systems. Cardiovascular research in the field of non-linear dynamics has mainly focused on heart rate, and less importance has been given to blood pressure variability, though continuous BP fluctuations are intrinsic characteristics of cardiovascular system (Papaioannou et al. 2006).

6.1.1 Blood Pressure

Blood pressure is another important physiological signal which is of great concern to health experts as there is no alternative method of measuring other than the sphygmomanometer. The blood pressure is thought to carry significant information regarding ones physical features. Blood pressure measurement can help to illustrate cardiovascular diseases, like hypertension, heart attack, and asthma. Thus blood pressure is an important biomarker of cardiovascular health (Klabunde 2005). As blood circulates in the arteries, it exerts pressure on the walls of the arteries which is noted as blood pressure (Bojanov 2005). It is the driving force which pushes the blood to flow in the vessels. Way back in 1733 blood pressure was first measured. Stephen Hales a veteran doctor used brass pipes to measure blood pressure in animals. In 1896, an Italian physician Scipione Riva-Rocci developed the instrument to measure blood pressure – the sphygmomanometer (Ward et al. 2007). Towards the end of the cardiac cycle when the ventricles contract, blood flows from the ventricles to the arteries and the maximum pressure in the arteries is measured as systolic blood pressure (SBP), whereas ventricles filled with blood at the beginning of the cardiac cycle exert minimum pressure on the arteries during ventricular diastole, noted as diastolic blood pressure (DBP) (Bojanov 2005). Normal blood pressure in adults is 120/80 mmHg where 120 mmHg is the systolic pressure and 80 mmHg is the diastolic pressure. Large deviation of these values is an indication of cardiac disease.

The analytical method for blood pressure or changes in heart rate is generally a linear process, but normal physiological modulations of BP and HR are considered to be mostly complex and non-linear. Several authors have advocated the fact that conventional time and frequency domain analysis techniques based on the linear fluctuation of heart rate are insufficient to outline the changes in heart rate dynamics (Coenen et al. 1977; Fakhouri 1980; Dasheiff and Dickinson 1986; Oppenheimer 2001; Berilgen et al. 2004; Lahrmann et al. 2006; Kamal 2006, 2010; De Ferrari et al. 2009; Foldvary-Schaefer and Unnwongse 2011; Zamponi et al. 2011; Meregnani et al. 2011); thus, new techniques in the domain of non-linear dynamics have been initiated to quantify complex heart rate dynamics and complement conventional measures of its variability (Kamal 2014). The principles of non-linear dynamics including chaos theory and fractal concepts have been investigated to provide a better understanding of the transition(s) that occur during the transition from a healthy to a pathological state (Goldberger et al. 1988). These authors emphasize that it is the degree of variability in say, for example, the heart rate that is a characteristic of a healthy individual. A gradual decrease in the variability indicates a transition from a healthy state to a pathological state. Analysis of heart failure is of tremendous significance since it is a major medical problem that affects the human population on a large scale worldwide.

Papaioannou et al. (2006) investigated the effect of caffeine on indices expressing the complex and "chaotic" non-linear characteristics of BP variability. Kinnane et al. (2003) analyzed BP signals with the help of different non-linear time series analysis techniques. The possible factors that could affect blood pressure were either removed or enhanced experimentally so that the mechanisms controlling BP could be identified by chaotic analysis of the signals. For different experimental conditions, the level of chaos varied demonstrating significant reduction in case of other experimental conditions compared to control conditions. Pavlov et al. (2005) discussed how stress affects the features of multifractality in the cardiovascular dynamics using wavelet transform modulus maxima method. For arterial blood pressure recordings in healthy rats, the study reported that the stress-induced changes of multifractality may be different for male and female organisms. They concluded that for male rats stress reduces "smoothness" of blood pressure dynamics and sometimes may also reduce the degree of multifractality, whereas female rats showed less sensitive to stresses. Some works have reported an increase in systolic and diastolic pressures during head-up tilt in normal subjects (Cooke et al. 1999; Porta et al. 2012). Gospodinova and Gospodinov (2014) investigated arterial vibrations of subjects with normal and high blood pressure using the fractal and multifractal methods. Using DFA method the authors defined the complexity of investigated data through significant differences in scalable behavior between healthy subjects with normal blood pressure and pathological cases with high blood pressure. On the base of the multifractal analysis, they concluded that the investigated signals, corresponding to the healthy subjects, showed multifractal behavior and in pathological cases the signals are monofractals. The fractal and multifractal analyses of the investigated signals corresponding with arterial vibrations of the people with normal and high blood pressure show that they are suitable for noninvasive methods of diagnostics, forecast, and prevention of the pathological statuses.

6.1.2 Non-linear Heart Rate Variability Analysis

We know that the electrical activity of heart cells is measured by the electrocardiogram signal. The impulse causes the rhythmic contraction of the heart, where the electrical performance of the heart is represented by beats (Estrada et al. 2014). Pioneer work of Glass and Mackey (1988) introduced non-linear approaches into heart rhythm analysis. Ritzenberg et al. (1984) were the first to provide evidence of non-linear behavior in the electrocardiogram and arterial blood pressure traces of a dog that were injected with noradrenaline. Goldberger and West (1987) were the first to analyze HR variability (HRV) using non-linear fractal dynamics. They opined that the "constrained randomness" observed for physiological variability and adaptability can be explained by fractal scale invariance. Goldberger et al. (1988) reported that patients with high risk of sudden cardiac death exhibit non-linear HR dynamics along with abrupt spectral changes and sustained low-frequency (LF) oscillations. Later in 1991 Goldberger suggested that due to certain pathological conditions like reduced HR dynamics, the complexity of physiological variability may lose prior to sudden death and aging (Goldberger 1991).

Multivariate non-linear analysis of HRV was performed for the first time by Babloyantz and Destexhe (1988). For quantification of non-linearity, ECG obtained from four normal subjects were analyzed qualitatively and quantitatively using phase portrait, Poincaré section, correlation dimension, Lyapunov exponent, and Kolmogorov entropy. They found variability underlying inter-beat intervals to be not random, but to reveal short-range correlations controlled by deterministic laws. Chaffin et al. (1991) applied phase space reconstruction and dimensional analysis to study HR traces recorded from scalp electrodes of 12 healthy fetuses. For estimating the complexity of cardiovascular system, Pincus (1991) amended the original version of correlation dimension and Kolmogorov entropy (Grassberger and Procaccia 1983; Eckmann and Ruelle 1985), to develop approximate entropy (ApEn). Richman and Moorman (2000) later upgraded approximate entropy (ApEn) to "sample entropy" (SampEn). Decrease in the value of SampEn in case of neonatal HR before clinical diagnosis of sepsis and sepsis-like illness was reported by Lake et al. (2002). Tuzcu et al. (2006) reported low values of SampEn prior to emergence of atrial fibrillation (Voss et al. 2009). Peng et al. (1995) applied DFA to characterize the fractal structure of HR. Ivanov et al. (1999) discovered multifractality in HR dynamics and showed that the heartbeat modulation requires multiple scaling exponents for its characterization. For determining complexity over multiple scales, Costa et al. (2002, 2005) presented new techniques. In deadly conditions loss of multiscale complexity (Norris et al. 2008) advocates clinical importance of multiscale complexity measure. Humeau et al. (2008) used wavelet-based representations, Holder exponents, and sample entropy to quantify laser Doppler flowmetry (LDF) signals which allow the monitoring of microvascular blood flow, thereby providing a peripheral view of the cardiovascular system. The results indicated a possible modification of the peripheral cardiovascular system with aging. Thus, the endothelial-related metabolic activity was found to decrease, but not significantly, with aging. Further the LDL signals were found to be greatly monofractal in elderly subjects compared to the younger subjects where LDF signals are weakly multifractal. The average mean sample entropy value of LDF signals was found to decrease slightly with age. Thus the authors concluded that the study can help in acquiring knowledge about the relationship between the status of microvascular system and age thus leading to age-related non-linear modeling which would be more authentic (Humeau et al. 2008). Tseng et al. (2013) used linear and non-linear techniques to study healthy subjects or subjects with postural hypotension, while subjects underwent the tilt table exam. The study was conducted in two phases. In the first phase, change in HR during tilt table test was analyzed using power spectral density (PSD) analysis, detrended fluctuation analysis (DFA), and multiscale entropy (MSE) analysis, and in the second phase, change between BP and CBFV was analyzed using MSE and correlation coefficient analysis (Tseng et al. 2013). Moga et al. (2014) conducted a study to determine the physiological basis of methods for computing the dynamics of beat-to-beat RR interval. New methods for estimating ventricular arrhythmia and sudden cardiac death in congestive heart failure patients were also determined (Malik and Camm 1995). Cornforth et al. (2015) compared three multiscale measures, namely, multiscale entropy (MSE), multifractal detrended fluctuation analysis (MFDFA), and Renyi entropy (RE), to compare data of RR intervals obtained from cardiac autonomic neuropathy (CAN) patients and aged controls. Castiglioni and Merati (2017) applied fractal analysis to study HR variability in paraplegic patients. Fractal analysis was found to be superior compared to traditional power spectral analysis in a subgroup of paraplegic subjects with sound cardiac health. In a recent study, Castiglioni et al. (2018) described the multifractal and multiscale characteristics of cardiovascular signals in healthy subjects under controlled conditions. Using DFA-based method, they compared three cardiovascular signals, namely, inter-beat interval (IBI, inverse of heart rate), SBP, and DBP signals, recorded in 42 female and 42 male volunteers. The method optimizes data splitting in blocks to reduce the DFA estimation variance and to evaluate scale coefficients with Taylor's expansion formulas and maps the scales from beat domains to temporal domains. The analysis showed that scale coefficients and degree of multifractality depend on the temporal scale, with marked differences between IBI, SBP, and DBP and with significant sex differences. Results may be interpreted considering the distinct physiological mechanisms regulating heart rate and blood pressure dynamics and the different autonomic profiles of males and females.

6.1.3 Correlation of Heart Rate and Blood Pressure Signals

In order to understand the dynamics of the cardiovascular system, analysis of both HR and BP signals simultaneously is important as there is strong coherence between HR variability and BP variability (Saul et al. 1991; Pagani et al. 1986).

Chau et al. (1993) estimated fractal dimension of beat-to-beat HR and BP of control subjects and diabetic patients and observed decrease in fractal dimension of HR in diabetic patients than in healthy subjects. From their study they concluded heartbeat to show greater fractal features in healthy condition, whereas a low fluctuation signifies pathology. Evidence of a close non-linear coupling between the respiratory and cardiovascular system was provided in 1993 by Novak et al. (1993). Several other authors using a variety (Voss et al. 2009) of non-linear techniques investigated the interactions and couplings between HR and respiration and HR and BP, respectively (Parati et al. 1988; Pompi et al. 1998; Baumert et al. 2002; Schwab et al. 2006; Fuchs et al. 2010). Some studies investigated spectral analysis of HRV in response to the posture changes maneuver and dynamic exercise

(Yamamoto and Hughson 1991; Butler et al. 1993; Nakamura et al. 1993; Ahmed et al. 1994; Montano et al. 1994; Radelli et al. 1994; Fei et al. 1995; Tulppo et al. 1996,1998; Tonhajzerov et al. 2002; Pichon et al. 2006). Hughson et al. (1995) explored harmonic and fractal components of BP variability in heart transplant patients and in age- and sex-matched healthy people while seated rest, supine rest, and supine rest with fixed-pace breathing of 12 respirations per minute. In transplant patients HR was found to be much faster compared to healthy controls. Compared to supine rest or the supine plus fixed breathing position, RR intervals of healthy controls were found to be distinctively shorter in upright seated position. In the transplant patients, in the upright seated position, RR intervals were found to be significantly shorter compared to in the supine rest position. In control subjects total power of systolic or diastolic pressure did not differ when BP variability was analyzed using spectral analysis. Systolic or diastolic pressure in transplant patients revealed less low-frequency harmonic spectral power and more high-frequency power in diastolic pressure compared to healthy controls. Compared to healthy controls, transplant patients showed persistently higher ratio of high-frequency power in diastolic relative to systolic pressure. Transplant patients and control subjects both revealed almost same slope of the fractal component of systolic pressure which was higher than the slope of HR variability (in control subjects) (Hughson et al. 1995). Shoemaker et al. (2001) conducted a study to test the hypothesis that sympathetic adjustments to tilt are attenuated in women versus men leading to diminished blood pressure responses to head-up tilt (HUT). Kuusela et al. (2002) conducted a study using different non-linear methods to characterize HR and BP dynamics in healthy subjects at rest. Castellano et al. (2004) measured blood flow in the carotid and femoral arteries, heart rate, and blood pressure in response to postural challenge in older adults. Tulppo et al. (2005) reported the behavior of the alpha-1 during the passive tilt test. The knowledge of physiological responses induced by autonomic tests is relevant to provide further information regarding autonomic cardiac regulation and autonomic dysfunction diagnosis (Ziegler 1994). Though the responses caused by the passive orthostatic test are elaborated in Tonhajzerov et al. (2002) and Pichon et al. (2006), it was not clear how long the autonomic nervous system spends to induce sympathetic and parasympathetic changes immediately after the change of position.

Špulák et al. (2010) in a study found the correlation coefficients between systolic BP and parameters computed from ECG and photoplethysmography (PPG) to vary strongly subject to subject. Corino et al. (2010) using a symbolic distance showed the short-term dynamics of systolic arterial pressure (SAP) to vary distinctly in rest and tilt phases, whereas RR dynamics was left unaltered. Pachauri and Mishra (2012) investigated the phase synchronization between ECG and arterial BP in order to find the interactions between the two signals. Gesche et al. (2012) in a work showed that the created pulse wave velocity (PWV) – BP function, including a one-point calibration – produced significant correlation between BP derived from the PWV and the systolic BP measured by sphygmomanometer. Ahmad et al. (2012) presented a method whereby ECG-assisted oscillometric and pulse transit time (PTT) analyses were seamlessly integrated into the oscillometric BP measurement paradigm. The method bolstered oscillometric analysis (amplitude modulation) with more reliable ECG peaks provided a complementary measure with PTT analysis (temporal modulation) and fused this information for robust BP estimation. Bishop et al. (2012) employed the wavelet modulus maxima technique to characterize the multifractal properties of HR and mean arterial BP physiology retrospectively for four patients during open abdominal aortic aneurysm repair. Faini et al. (2013) described changes in day and night fractal dimension (FD) of BP and HR of a large population. HR and systolic and diastolic BP time series of hypertensive subjects were obtained from ambulatory blood pressure monitoring (ABPM) during the day and at night. FD was calculated by Higuchi's algorithm (FD_H) and Katz's corrected algorithm (FD_c). Standard deviation (SD) was found to decrease significantly from day to night for systolic BP (SBP), diastolic BP (DBP), and HR, while coefficient of variation decreased significantly for HR only. Whereas same FD was noted for SBP and DBP during the day, while at night changes occurred for DBP only. Changes in the regulation of vascular resistances at night probably associated with lying position are represented by the existence of night–day differences only in the FD of DBP. Further the authors also noted alike trends with FD estimators based on different algorithms (FD_H and FD_c). In another study Faini et al. (2014) quantified mean, SD, and FD at daytime and nighttime of 47 normotensive male volunteers. As expected they observed mean and SD to decrease from day to night, whereas FD of SBP increased significantly than SBP at night. FD of HR was observed to be distinctively lower compared to FD of SBP or DBP. Thus they concluded that fractal dimension of ambulatory BP may present modern and cutting-edge information of cardiovascular system. Estrada et al. (2014) with the use of neural networks demonstrated the existence of a relationship between ECG signals and BP. Souza et al. (2014) aimed to investigate the effects of the posture changes maneuver on fractal exponents through DFA in young women, as well as the time and frequency domains indexes of HR. Pujitha et al. (2014) investigated the postural changes in HR and BP with aging. They concluded that due to stiffening of blood vessels, BP increase with increase in age, but the postural decrease in systolic BP in standing from lying down posture was more in elderly subjects. Several other researchers have also investigated the relationship between postural changes and cardiovascular response via the HR and BP (Ewing et al. 1980; Borst et al. 1982; Pump et al. 1997, 1999, 2001).

Though a very few studies on correlation between ECG and arterial BP (ABP) due to change in posture have been reported earlier as illustrated in literature, to the best of our knowledge, no effort has been made to see if there exists any cross-correlation between the two signals. Heart rate variability and blood pressure variability are referred to as being capable of foretelling cardiovascular risks (Dawson et al. 2000); hence precise measurement and diagnosis of these parameters are required to avoid mortality. Thus with a motive to understand the correlation between the two signals, we used a non-linear cross-correlation technique called multifractal detrended cross-correlation analysis (MF-DXA) proposed by Zhou (2008) to assess the time series of ECG and ABP. The detail of the method is provided in Appendix B.

6.2 Posture-Dependent ECG and Arterial Blood Pressure (ABP) Data

To explore the cross-correlation between ECG and arterial BP (ABP), we studied the database of an experiment from website www.physionet.org (www.physionet.org/physiobank/database/prcp/) for 10 s. The database contains data of ABP and ECG collected from ten healthy subjects at rest, during rapid tilts, slow tilts, and stand tests. Out of ten healthy subjects, we have chosen eight healthy subjects for our study. The mean age of the subjects was 28.7 ± 1.2 years, the mean height was 172.8 ± 4.0 cm, and the mean weight was 70.6 ± 4.5 kg (Heldt et al. 2003). Participants regularly engaged in light to moderate physical activity and had no sign of cardiovascular disease. Plot of each of the two signals of ECG and ABP for 2 s is shown in Fig. 6.1a, b.

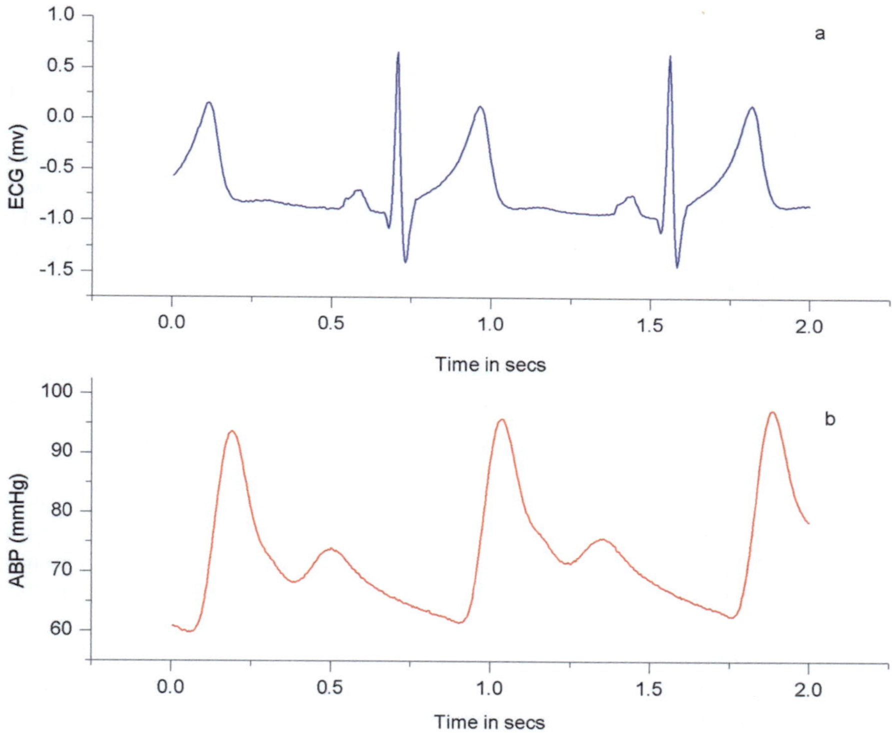

Fig. 6.1 Plot of (**a**) ECG and (**b**) ABP signal for a particular subject for 2 s (Ghosh et al. 2018)

6.3 Multifractal Cross-Correlation Analysis Between ECG and ABP Data

Multifractal detrended cross-correlation analysis technique can reveal multifractal characteristics of two cross-correlated signals and higher-dimensional multifractal measures. We have applied this technique in our previous works to study the EEG pattern of epileptic patients (Ghosh et al. 2014) and also to study the human gait pattern of normal people and patients suffering from Parkinson's disease (Dutta et al. 2016).

There are various other recently proposed methods like multifractal cross-correlation analysis (MF-CCA) used for studying the cross-correlation between two series. Kwapien et al. (2015) proposed a q-dependent detrended cross-correlation coefficient ρ_q, based on the q-dependent fluctuation functions F_q from MF-DFA and MF-DCCA (Kantelhardt et al. 2002; Oswiecimka et al. 2014), and showed that the new coefficient is not only able to quantify the strength of correlations but also allows one to identify the range of detrended fluctuation amplitudes in two signals (Kwapien et al. 2015). Qian et al. (2015) recently used detrended partial cross-correlation analysis (DPXA) to uncover the intrinsic power-law cross-correlations between two simultaneously recorded time series in the presence of non-stationarity after removing the effects of other time series acting as common forces. They analyzed the multifractal binomial measures masked with strong white noises using multifractal DPXA (MF-DPXA) and found that the MF-DPXA method quantifies the hidden multifractal nature which the MF-DCCA method fails to do. In another work Jiang et al. (2017a) proposed a method for characterizing the joint multifractal nature of long-range cross-correlations and named it joint multifractal analysis based on wavelet leaders (MF-X-WL). They found the MF-X-WL method can detect, respectively, the joint multifractality and monofractality in binomial measures and bivariate fractional Brownian motions, but not with accuracy. Jiang et al. (2017b) proposed another method called multifractal cross wavelet analysis (MF-X-WT) which is based on wavelet analysis and is used to characterize the joint multifractal nature of long-range cross-correlations. Similar to the MF-X-PF method, the authors introduced two orders in the MF-X-WT (p, q) method and assessed its performance by conducting extensive numerical experiments on the dual binomial measures with multifractal cross-correlations and the bivariate fractional Brownian motions (bFBMs) with monofractal cross-correlations. On applying the method to stock market indexes, intriguing joint multifractal nature was observed in pairs of index returns and volatilities. Since these recently proposed methods are yet to be applied in different domains, we used MF-DXA as several studies conducted so far using this technique have proved the robustness of this method.

Multifractal analysis was employed for both ECG and ABP time series. Both the time series were first transformed to obtain the integrated signal and reduce noise in the data. The integrated time series was then divided into N_s bins where $N_s = \text{int}(N/s)$, N is the length of the series, and s is the length of the bin. For both the signals, the range of s was chosen from 10 to $N/10$ in steps of 1. The qth order fluctuation

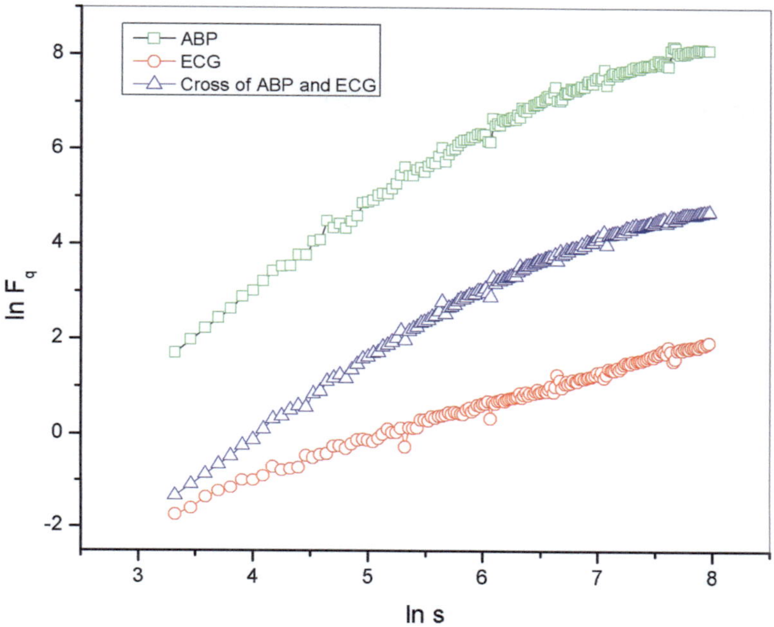

Fig. 6.2 Plot of ln F_q vs. ln s ($q = 2$) for a particular subject (Ghosh et al. 2018)

function $F_q(s)$ was determined for $q = -10$ to $+10$ in steps of 1. Plot of ln $F_q(s)$ vs. ln s for a particular subject for both auto- and cross-correlated series of ECG and ABP is depicted in Fig. 6.2. Power-law scaling of the fluctuation function ($F_q(s)$ vs. s) is observed for all values of q. Linear dependence of ln F_q on ln s for different values of q suggests that there exists power-law auto- and cross-correlation between ECG and ABP signals. The values of cross-correlation scaling exponent $\lambda(q)$ are obtained from the slope of linear fit of ln $F_q(s)$ vs. ln s plots. The values of Hurst exponent h ($q = 2$) and cross-correlation scaling exponent $\lambda(q = 2)$ are shown in Table 6.1. Figure 6.3 shows the relationship between scaling exponent $\lambda(q)$ and q for a particular subject. For comparison the relation between $h(q)$ and q obtained using MF-DFA technique is also shown in the same figure. Dependence of both Hurst exponent and scaling exponent on q suggests multifractal behavior, i.e., there are different power-law auto- and cross-correlations. The dependence of the generalized Hurst exponents $h(q)$, and scaling exponents, $\lambda(q)$, suggests different scaling of small and large fluctuations, i.e., the scaling is multifractal. From Fig. 6.3 we can also see that the value of $\lambda(q)$ for $q < 0$ is larger than that of $q > 0$ for the original series, whereas for the shuffled series, the exponent is constant. Figure 6.3 portrays the relation where the cross-correction scaling exponent $\lambda(q)$ is equal to the average of the generalized Hurst exponent $h(q)$, i.e., $\lambda(q = 2) \approx [h_x(q = 2)+h_y(q = 2)]/2$. The equation is found to be approximately valid in all the cases within the limit of experimental error. For artificially generated time series, the above relation is valid

Table 6.1 Values of $h(q = 2)$ for ECG and ABP, cross-correlation scaling exponent (λ), auto-correlation (γ), and cross-correlation (γ_x) coefficients along with mean and standard deviation for ECG and ABP (Ghosh et al. 2018)

Subject	$h(q = 2)$ ECG	$h(q = 2)$ ABP	λ	γ (ECG)	γ (ABP)	(γ_x)
1	0.751 ± 0.005	1.421 ± 0.014	1.334 ± 0.014	0.5 ± 0.010	−0.84 ± 0.028	−0.668 ± 0.020
2	0.684 ± 0.008	1.274 ± 0.013	1.297 ± 0.015	0.632 ± 0.016	−0.548 ± 0.026	−0.594 ± 0.032
3	1.042 ± 0.004	1.478 ± 0.014	1.385 ± 0.011	−0.084 ± 0.008	−0.956 ± 0.028	−0.768 ± 0.022
4	0.774 ± 0.007	1.405 ± 0.016	1.306 ± 0.014	0.452 ± 0.014	−0.809 ± 0.032	−0.612 ± 0.028
5	0.758 ± 0.006	1.431 ± 0.015	1.387 ± 0.012	0.484 ± 0.012	−0.862 ± 0.030	−0.772 ± 0.024
6	1.005 ± 0.004	1.432 ± 0.014	1.433 ± 0.011	−0.01 ± 0.008	−0.864 ± 0.028	−0.866 ± 0.022
7	0.798 ± 0.007	1.421 ± 0.013	1.359 ± 0.013	0.404 ± 0.014	−0.842 ± 0.026	−0.718 ± 0.026
8	0.788 ± 0.006	1.431 ± 0.013	1.425 ± 0.010	0.424 ± 0.012	−0.861 ± 0.026	−0.848 ± 0.020
Mean and SD	0.825 ± 0.128	1.412 ± 0.060	1.366 ± 0.051	0.483 ± 0.081	−0.823 ± 0.119	−0.731 ± 0.101

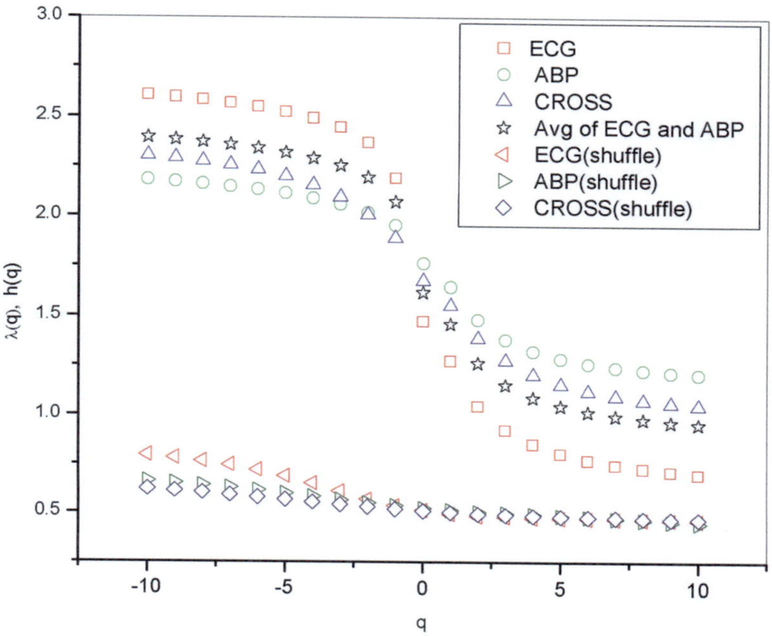

Fig. 6.3 Plot of $h(q)$ and $\lambda(q)$ vs. q of original and shuffled series for a particular subject (Ghosh et al. 2018)

for any value of q. However the relation is not valid for any value of q especially for the negative values of q. In Fig. 6.3 we have also shown the values of bivariate Hurst exponent H_{xy} ($= (h_x + h_y)/2$), i.e., the average of generalized Hurst exponents of ECG and ABP (Ghosh et al. 2018). Oswiecimka et al. (2014) have shown that the more the difference between $\lambda(q)$ and average generalized Hurst exponent, the more different are the considered multifractals.

In a recent work, Kristoufek (2015a) opined that the bivariate Hurst exponent H_{xy} is not necessarily equal to the average of the separate Hurst exponents. He argued that unless at least one of the series is long-range correlated with h (Hurst exponent) > 0.5, the processes cannot be power-law cross-correlated with $H_{xy} > 0.5$. Thus it is essential for one of the underlying processes to have long-term memory. The power-law cross-correlations are thus a by-product of the persistent separate processes. Ref (Podobnik et al. 2011; Kristoufek 2013, 2015b) describes long-range cross-correlated processes with analytical results. The author has also argued about another possibility where $H_{xy} < 1/2\ (h_x + h_y)$. Sela and Hurvich (2012) referred to such processes as the anti-cointegration as the separate processes are long-range correlated but pairwise uncorrelated in a long-term horizon (at low frequencies). In another work Kristoufek (2016) has given an explanation for the frequently reported estimated bivariate Hurst exponent H_{xy} being higher than the average of the separate Hurst exponents h_x and h_y, i.e., $H_{xy} > 1/2\ (h_x + h_y)$. As the Hurst exponent estimation

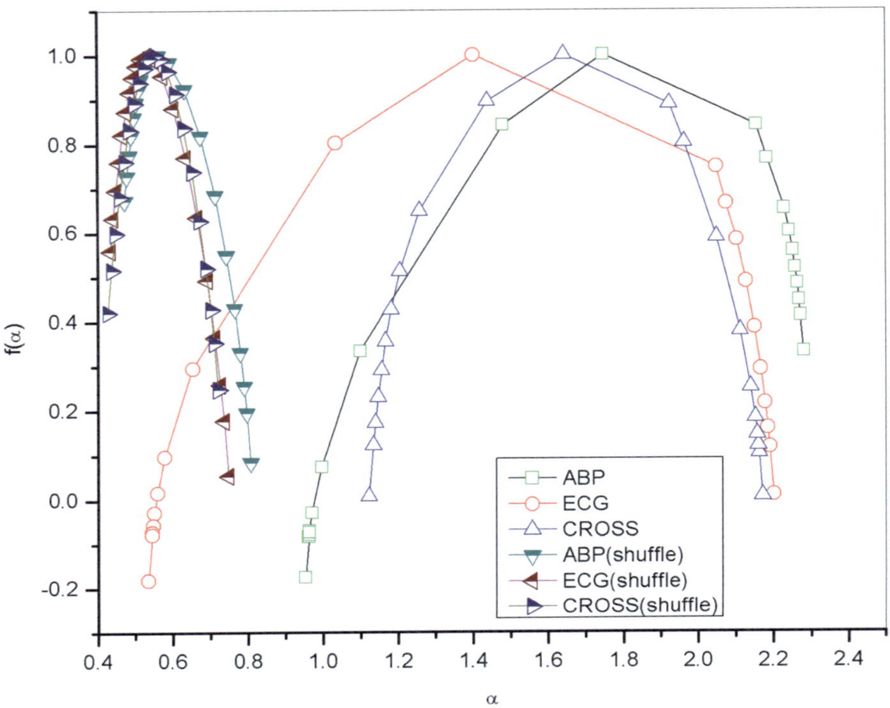

Fig. 6.4 Plot of $f(\alpha)$ vs. α of original and shuffled series for a particular subject (Ghosh et al. 2018)

is unbiased by heavy tails in the univariate setting, the upward bias in the bivariate setting for heavy tails implies the possibility of $H_{xy} > 1/2\ (h_x + h_y)$. The author also argues that one of the possible reasons for reporting $H_{xy} > 1/2\ (h_x + h_y)$ is a finite sample bias (Ghosh et al. 2018).

Figure 6.4 depicts the plot of $f(\alpha)$ vs. α. From the figure we can see that for $q = 2$ the cross-correlation scaling exponent $\lambda(q)$ is greater than 0.5 which is an indication that persistent long-range cross-correlation exists among ECG and ABP signals. The values of $h(q)$ for $q = 2$ for ECG and ABP are also observed to be greater than 0.5 which also speaks about persistent long-range correlation of the two signals (Ghosh et al. 2018). This phenomenon has also been observed in our previous studies (Ghosh et al. 2014; Dutta et al. 2016) as well.

According to the MF-DXA, we consider that each of the two variables at any time depends not only on its own past values but also on past values of the other variable (Wang et al. 2013). Thus the existence of power-law cross-correlations among ECG and ABP suggests that with change in posture, a change in the value of one is expected to create a subjective influence in change in value of the other. We further confirm that long-range correlation plays a dominant part in the existence of multifractal features in ECG and ABP (Ghosh et al. 2018).

Values of auto-correlation (γ) and cross-correlation (γ_x) for ECG and ABP were estimated and are presented in Table 6.1. The mean along with standard deviation is also shown in the same table. A lower value of γ indicates a higher degree of correlation. We have obtained negative values of cross-correlation (γ_x) for all subjects. Drozdz et al. (2009) have also observed negative values of correlation. Jones and Kaul (1996) were the first to reveal a stable negative cross-correlation between oil prices and stock prices. The negative cross-correlations were also found by (Chen 2010; Filis 2010; Berument et al. 2010). It is observed that for Subject 2 both the degree of auto-correlation and cross-correlation among ECG and ABP is the least compared to other subjects which may be an indication of complex cardiac condition in different posture. From Table 6.1 we further observe strong auto-correlation of ECG and ABP signals for Subject 3 (Ghosh et al. 2018). Significant cross-correlation between ECG and ABP is also observed for this subject which is in accordance with findings of Podobnik et al. (2007, 2009) as they showed that if the auto-correlations are stronger, then the cross-correlations will also be stronger. Table 6.1 also shows least value of cross-correlation (γ_x) for Subject 6 indicating strong cross-correlation between ECG and ABP implying healthy heart condition. Table 6.2 depicts values of multifractal width of ECG, ABP, and cross-correlated series. Except for Subject 3 the cross-correlated series of all other subjects shows weaker multifractality than ECG or ABP. The plot of multifractal spectrum for ECG, ABP, and cross-correlation series for a particular subject is shown in Fig. 6.4. Weaker multifractality is observed for the cross-correlated series (Ghosh et al. 2018). Our earlier studies have also reported this phenomenon (Ghosh et al. 2014; Dutta et al. 2016). Wang and Xie (2012) have also observed this feature. He and Chen also found especially, for the soy meal, soybean, and corn futures markets, the widths of cross-correlation multifractal spectra to be narrower than those of separately analyzed China's and US soy meal futures markets using the MF-DXA method (He and Chen 2011). Various studies have reported weaker multifractality of the cross-correlated signal than the individual signals (Movahed and Hermanis 2008; Zhao et al. 2011). From Fig. 6.5 we observe the non-linear dependence of classical scaling exponent on q which is another piece of evidence of multifractality of the original and cross-correlated ECG and ABP signals.

Further it is very interesting to observe that Subject 2 having highest value of multifractal width (w) of ECG as shown in Table 6.2 shows lowest degree of cross-correlation among ECG and ABP pattern (as evident from Table 6.1) which may serve as a double check on dysfunction of cardiovascular system. Since larger value of multifractal width (w) is a measure of greater complexity, hence value of w of ECG in case of Subject 2 may provide an indication of cardiovascular health dysfunction. Dutta (2010) have also reported larger values of multifractal width of ECG signals of diseased subjects compared to healthy ones. To ascertain the origin of multifractality of both the ECG and ABP time series, we randomly shuffled the series and then analyzed them. Figures 6.3 and 6.4, respectively, depict plots of $h(q)$, $\lambda(q)$ vs. q, $f(\alpha)$ vs. α for the original and randomly shuffled series. Both the figures depict weaker multifractality of the shuffled series (Ghosh et al. 2018).

Table 6.2 Values of multifractal width (w) of ECG, ABP, and cross-correlation (w_x) of original and shuffled series (Ghosh et al. 2018)

Subject	w(original) ECG	w(shuffle) ECG	w(original) ABP	w(shuffle) ABP	w_x(original) Cross	w_x(shuffle) Cross
1	2.401 ± 0.198	0.940 ± 0.030	1.365 ± 0.070	0.563 ± 0.016	1.404 ± 0.062	0.558 ± 0.012
2	2.465 ± 0.145	0.653 ± 0.054	1.650 ± 0.085	0.519 ± 0.072	1.481 ± 0.051	0.437 ± 0.018
3	2.176 ± 0.163	0.677 ± 0.035	1.283 ± 0.092	0.424 ± 0.011	1.429 ± 0.051	0.390 ± 0.014
4	1.726 ± 0.103	0.724 ± 0.015	2.378 ± 0.073	0.622 ± 0.022	1.545 ± 0.043	0.622 ± 0.022
5	2.086 ± 0.161	0.640 ± 0.018	1.407 ± 0.070	0.505 ± 0.043	1.156 ± 0.037	0.348 ± 0.021
6	1.725 ± 0.126	0.396 ± 0.013	1.489 ± 0.084	0.463 ± 0.019	1.110 ± 0.050	0.363 ± 0.005
7	2.016 ± 0.382	1.021 ± 0.040	1.384 ± 0.085	0.479 ± 0.020	1.203 ± 0.124	0.584 ± 0.013
8	2.192 ± 0.273	1.088 ± 0.048	1.411 ± 0.140	0.444 ± 0.009	1.240 ± 0.095	0.586 ± 0.015

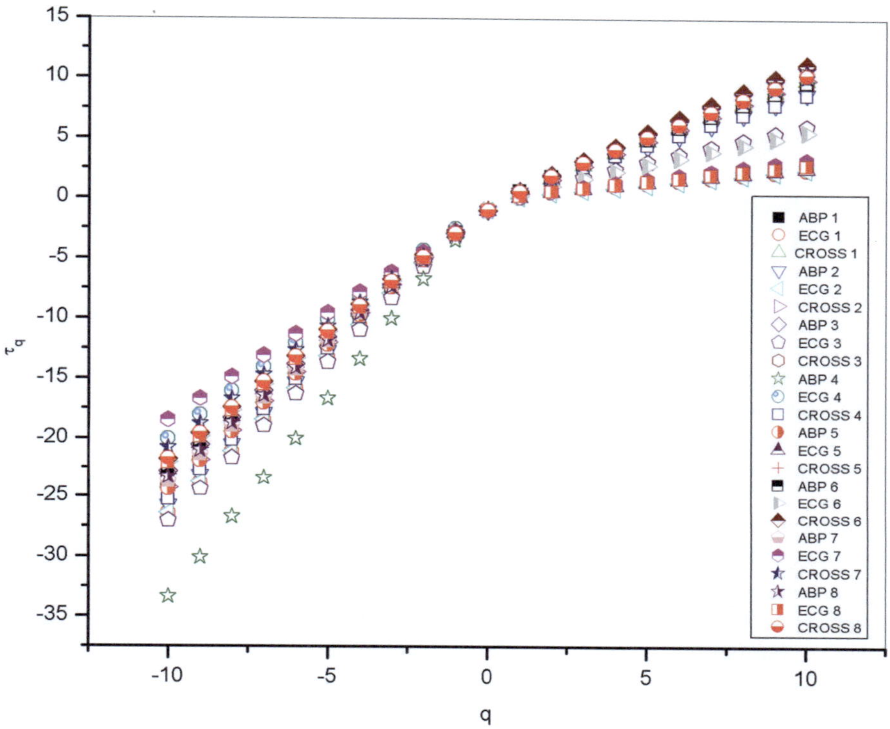

Fig. 6.5 Plot of $\tau(q)$ vs. q for eight subjects (Ghosh et al. 2018)

Table 6.3 Values of auto-correlation (γ) and cross-correlation (γ_x) coefficients for ECG and ABP of shuffled series

Subject	γ (ECG) shuffle	γ (ABP) shuffle	(γ_x) shuffle
1	0.948 ± 0.005	0.976 ± 0.004	0.902 ± 0.003
2	0.896 ± 0.007	0.913 ± 0.006	0.871 ± 0.004
3	1.036 ± 0.004	0.994 ± 0.004	1.016 ± 0.003
4	0.970 ± 0.004	1.063 ± 0.004	0.974 ± 0.004
5	0.965 ± 0.005	0.913 ± 0.006	0.894 ± 0.006
6	0.963 ± 0.005	0.920 ± 0.004	0.923 ± 0.003
7	1.037 ± 0.005	1.009 ± 0.004	0.971 ± 0.004
8	1.079 ± 0.005	0.994 ± 0.005	0.902 ± 0.006

Table 6.2 clearly depicts the difference in values of multifractal width of the original and shuffled series. We observe weaker multifractality for the shuffled series which implies that both types of multifractality are present, but the more dominant factor is long-range correlations. Figure 6.5 provides evidence of another piece of multifractality where we can see non-linear variation of $\tau(q)$ vs. q for all the eight subjects. We further observe from Table 6.3 that all the values of auto-correlation (γ)

and cross- correlation (γ_x) exponent for the shuffled series is close to 1, indicating all correlations are destroyed in the shuffling procedure as a value of 1 is indicative of uncorrelated data (Ghosh et al. 2018).

6.4 Discussion of the Result and Possible Use as Biomarker of Neurological Disorder

Thus the above analysis of cross-correlation among ECG and ABP in healthy subjects using MF-DXA methodology reveals important and interesting observations. This analysis clearly indicates that the change of posture gives rise to autonomic deregulation. This correlation between autonomic deregulation and ECG is of course subjective; in some cases the impact of change of posture is compatibly higher. Another important feature is that the subject having the lowest impact possesses more complex ECG pattern as evident from multifractal widths. It should be noted that the subject chosen were all young having age around 28 years. Since these types of data of older people are not available, we are not in a position to assess age-dependent impact of posture. Nevertheless this data will serve as basis of comparison of impact at older age which in future should be a subject of great concern. Further this analysis may be platform to develop a sensitive and rigorous biomarker for assessment of cardiovascular responses to change in posture inducing autonomic deregulation. Appropriate user-friendly electronic gadget may be developed and used extensively. The possibility of using this method in the domain of orthostatic syndrome seen post flight is another interesting area of research. Finally this work presents for the first time information about heart condition in different posture from the perspective of chaos and fractality yielding results of high precision.

Acknowledgment The authors gratefully acknowledge Physica A and Elsevier Publishing Co. for providing the copyrights of Figs. 6.1a, b, 6.2, 6.3, 6.4, and 6.5 and Tables 6.1, 6.2, and 6.3.

References

Acharya UR, Joseph KP, Kannathal N, Lim CM, Suri JS (2006) Heart rate variability: a review. Med Biol Eng Comput 44:1031–1051

Ahmad S, Chen S, Soueidan K, Batkin I, Bolic M et al (2012) Electrocardiogram – assisted blood pressure estimation. IEEE Trans Biomed Eng 59:608–618

Ahmed MW, Kadish AH, Parker MA, Goldberger JJ (1994) Effect of physiologic and pharmacologic adrenergic stimulation on heart rate variability. J Am Coll Cardiol 24:1082–1090

Aubert AE, Ramaekers D (1999) Neurocardiology: the benefits of irregularity: the basics of methodology, physiology and currentclinical applications. Acta Cardiol 54:107–120

Aubert AE, Seps B, Beckers F (2003) Heart rate variability in athletes. Sports Med 33:889–919

Babloyantz A, Destexhe A (1988) Is the normal heart a periodic oscillator. Biol Cybern 58:203–211

Baumert M, Walther T, Hopfe J, Stepan H, Faber R et al (2002) Joint symbolic dynamic analysis of beat-to-beat interactions of heart rate and systolic blood pressure in normal pregnancy. Med Biol Eng Comput 40:241–245

Berilgen MS, Sari T, Bulut S, Mungen B (2004) Effects of epilepsy on autonomic nervous system and respiratory function tests. Epilepsy Behav 5:513–516

Berument MH, Ceylan N, Dogan N (2010) The impact of oil price shocks on the economic growth of selected MENA1 countries. Energy J 31:149–176

Bishop SM, Yarham SI, Navapurkar VU, Menon DK, Ercole A (2012) Multifractal analysis of hemodynamic behaviour: intraoperative instability and its pharmacological manipulation. Anesthesiology 117:810–821

Bojanov G (2005) In: Iaizzo PA (ed) Handbook of cardiac anatomy, physiology, and devices. Humana Press, Totowa

Borst C, Wieling W, Van Brederode JF, Hond A, de Rijk LG et al (1982) Mechanisms of initial heart rate response to postural change. Am J Phys 243:H676–H681

Butler GC, Yamamoto Y, Xing HC, Northey DR, Hughson RL (1993) Heart rate variability and fractal dimension during orthostatic challenges. J Appl Physiol 75:2602–2612

Castellano V, Olive JL, Stoner L, Black C, McCully KK (2004) Blood flow response to a postural challenge in older men and women. Dyn Med 3:1

Castiglioni P, Merati G (2017) Fractal analysis of heart rate variability reveals alterations of the integrative autonomic control of circulation in paraplegic individuals. Physiol Meas 38:774

Castiglioni P, Lazzeroni D, Coruzzi P, Faini A (2018) Multifractal-multiscale analysis of cardiovascular signals: a DFA-based characterization of blood pressure and heart-rate complexity by gender. Complexity, 2018, Article ID 4801924

Chaffin DG, Goldberg CC, Reed KL (1991) The dimension of chaos in the fetal heart rate. Am J Obstet Gynecol 165:1425–1429

Chau NP, Chanudet X, Bauduceau B, Gautier D, Larroque P (1993) Fractal dimension of heart rate and blood pressure in healthy subjects and in diabetic subjects. Blood Press 2:101–107

Chen SS (2010) Do higher oil prices push the stock market into bear territory? Energy Econ 32:490–495

Coenen AJ, Rompelman O, Kitney RI (1977) Measurement of heart-rate variability: part 2-hardware digital device for the assessment of heart-rate variability. Med Biol Eng Comput 15:423–430

Cooke WH, Hoag JB, Crossman AA, Kuusela TA, Tahvanainen KUO et al (1999) Human responses to upright tilt: a window on central autonomic integration. J Physiol 517:617–628

Corino VDA, Belletti S, Terranova P, Lombardi F, Mainardi LT (2010) Heart rate and systolic blood pressure in patients with persistent atrial fibrillation: a linguistic analysis. Methods Inf Med 49:516–520

Cornforth D, Jelinek HF, Tarvainen M (2015) A comparison of non-linear measures for the detection of cardiac autonomic neuropathy from heart rate variability. Entropy 17:1425–1440

Costa M, Goldberger AL, Peng CK (2002) Multiscale entropy analysis of complex physiologic time series. Phys Rev Lett 89:068–102

Costa M, Goldberger AL, Peng C-K (2005) Multiscale entropy analysis of biological signals. Phys Rev E 71:021–906

Dasheiff RM, Dickinson LJ (1986) Sudden unexpected death of epileptic patient due to cardiac arrhythmia after seizure. Arch Neurol 43:194–196

Dawson SL, Manktelow BN, Robinson TG, Panerai RB, Potter JF (2000) Which parameters of beat-to-beat blood pressure and variability best predict early outcome after acute ischemic stroke? Stroke 31:463–468

De Ferrari GM, Sanzo A, Schwartz PJ (2009) Chronic vagal stimulation in patients with congestive heart failure. In: Conference proceedings IEEE engineering in medicine and biology society, pp 2037–2039

Drozdz D, Kwapień J, Oświecimka P, Rak R (2009) Quantitative features of multifractal subtleties in time series. Europhys Lett 88:60003

Dutta S (2010) Multifractal properties of ECG patterns of patients suffering from congestive heart failure. J Stat Mech 2010:P12021

Dutta S, Ghosh D, Samanta S (2016) Non linear approach to study the dynamics of neurodegenerative diseases by multifractal detrended cross-correlation analysis – a quantitative assessment on gait disease. Physica A 448:181–195

Eckmann JP, Ruelle D (1985) Ergodic theory of chaos and strange attractors. Rev Mod Phys 57:617–656

Estrada GM, Mendoza LE, Molina V (2014) Relationship of blood pressure with the electrical signal of the heart using signal processing. Tecciencia 9:9–14

Ewing DJ, Hume L, Campbell IW, Murray A, Neilson JM et al (1980) Autonomic mechanisms in the initial heart rate response to standing. J Appl Physiol 49:809–814

Faini A, Parati G, Rienzo MD, Castiglioni P (2013) Night and day changes in heart rate and blood pressure fractal dimensions from 24-hour ambulatory blood pressure monitoring devices. Comput Cardiol 40:475–478

Faini A, Parati G, Bilo G, Rienzo MD, Castiglioni P (2014) Fractal characteristics of blood pressure and heart rate from ambulatory blood pressure monitored over 24 hours. In: 8th conference of the European Study Group on Cardiovascular Oscillations (ESGCO 2014), pp 73–74

Fakhouri SY (1980) Identification of Volterra kernels of non-linear discrete system. In: IEEE proceedings D control theory and applications, vol 127, pp 296–304

Fei L, Anderson MH, Statters DJ, Malik M, Camm AJ (1995) Effects of passive tilt and submaximal exercise on spectral heart rate variability in ventricular fibrillation patients without significant structural heart rate disease. Am Heart J 129:285–290

Filis G (2010) Macro economy, stock market and oil prices: do meaningful relationships exist among their cyclical fluctuations? Energy Econ 32:877–886

Foldvary-Schaefer N, Unnwongse K (2011) Localizing and lateralizing features of auras and seizures. Epilepsy Behav 20:160–166

Fuchs K, Schumann AY, Kuhnhold A, Guzik P, Piskorski J et al (2010) Comparing analysis of heart rate and blood pressure fluctuations in healthy subjects In: Proceedings of the 6th ESGCO 2010, April 12–14, Berlin Germany

Gesche H, Grosskurth D, Küchler G, Patzak A (2012) Continuous blood pressure measurement by using the pulse transit time: comparison to a cuff-based method. Eur J Appl Physiol 112:309–315

Ghosh D, Dutta S, Chakraborty S (2014) Multifractal detrended cross-correlation analysis for epileptic patient in seizure and seizure free status. Chaos, Solitons Fractals 67:1–10

Ghosh D, Dutta S, Chakraborty S, Samanta S (2018) Chaos based non-linear analysis to study cardiovascular responses to changes in posture. Physica A 512:392–403

Glass L, Mackey MC (1988) From clocks to chaos: the rhythms of life. Princeton University Press, Princeton

Goldberger AL (1991) Is the normal heartbeat chaotic or homeostatic? News Physiol Sci 6:87–91

Goldberger AL, West BJ (1987) Applications of non-linear dynamics to clinical cardiology. Ann NY Acad Sci 504:195–213

Goldberger AL, Rigney DR, Mietus J, Antman EM, Greenwald S (1988) Non-linear dynamics in sudden cardiac death syndrome: heart rate oscillations and bifurcations. Experientia 44:983–987

Gospodinova E, Gospodinov M (2014) Fractal and multifractal analysis of medical data. Comparative Analysis of Biological Signals

Grassberger P, Procaccia I (1983) Measuring the strangeness of strange attractors. Physica D 9:189–208

Hainsworth R (1998) Physiology of the cardiac autonomic system. In: Malik M (ed) Clinical guide to cardiac autonomic tests. Kluwer Academic Publishers, Dordrecht

He LY, Chen SP (2011) Multifractal detrended cross-correlation analysis of agricultural futures markets. Chaos, Solitons Fractals 44:355–361

Heldt T, Oefinger MB, Hoshiyama M, Mark RG (2003) Circulatory response to passive and active changes in posture. Comput Cardiol 30:263–266

Hughson RL, Maillet A, Dureau G, Yamamoto Y, Gharib C (1995) Spectral analysis of blood pressure variability in heart transplant patients. Hypertension 25:643–650

Humeau A, Blondeau FC, Rousseau D, Rousseau P, Trzepizur W et al (2008) Multifractality, sample entropy, and wavelet analyses for age-related changes in the peripheral cardiovascular system. Preliminary results. Am Assoc Physicists Med 35:717–723

Ivanov PC, Amaral LAN, Goldberger AL, Havlin S, Rosenblum MG et al (1999) Multifractality in human heartbeat dynamics. Nature 399:461–465

Jiang ZQ, Yang YH, Wang GJ, Zhou WX (2017a) Joint multifractal analysis based on wavelet leaders. Front Phys 12:128907

Jiang ZQ, Gao XL, Zhou WX, Stanley HE (2017b) Multifractal cross wavelet analysis. Fractals 25:1750054

Jones CM, Kaul G (1996) Oil and the stock markets. J Financ 51:463–491

Jones AYM, Kam C, Lai KW, Lee HY, Chow HT et al (2003) Changes in heart rate and R-wave amplitude with posture. Chin J Phys 46:63–69

Kamal A (2006) Assessment of autonomic function for healthy and diabetic patients using entrainment methods and spectral techniques. In: IEEE 32nd annual Northeast bioengineering conference, Easton, Pennsylvania, pp 161–162

Kamal AK (2010) Assessment of autonomic function in epileptic patients. Neurosciences (Riyadh) 15:244–248

Kamal AK (2014) Novel method for assessment of autonomic function in health and disease: an application to epilepsy. Int J Neurorehabil 1:133

Kantelhardt JW, Zschiegner SA, Koscielny-Bunde E, Havlin S, Bunde A et al (2002) Multifractal detrended fluctuation analysis of nonstationary signals. Physica A 316:87–114

Kinnane OP, Ringwood JV, Ramchandra R, Barrett CJ, Guild SJ et al (2003) Deterministic chaos in blood pressure signals during different physiological conditions. In: IFAC modelling and control in biomedical systems, Melbourne, Australia, 2003, pp 323–328

Klabunde RE (2005) Cardiovascular physiology concepts. Williams & Wilkins, Baltimore, p 2005

Kristoufek L (2013) Mixed-correlated ARFIMA processes for power-law cross-correlations. Physica A 392:6484–6493

Kristoufek L (2015a) Can the bivariate Hurst exponent be higher than an average of the separate Hurst exponents? Physica A 431:124–127

Kristoufek L (2015b) On the interplay between short and long term memory in the power-law cross-correlations setting. Physica A 421:218–222

Kristoufek L (2016) Power-law cross-correlations estimation under heavy tails. Commun Nonlinear Sci Numer Simul 40:163–172

Kuusela TA, Jartti TT, Tahvanainen KUO, Kaila TJ (2002) Non-linear methods of biosignal analysis in assessing terbutaline-induced heart rate and blood pressure changes. Am J Physiol Heart Circ Physiol 282:H773–H781

Kwapien J, Oswiecimka P, Drozdz D (2015) Detrended fluctuation analysis made flexible to detect range of cross-correlated fluctuations. Phys Rev E 92:052815

Lahrmann H, Cortelli P, Hilz M, Mathias CJ, Struhal W et al (2006) EFNS guidelines on the diagnosis and management of orthostatic hypotension. Eur J Neurol 13:930–936

Lake DE, Richman JS, Griffin MP, Moorman JR (2002) Sample entropy analysis of neonatal heart rate variabilityAmerican. Am J Phys Regul Integr Comp Phys 283:R789–R797

Malik M, Camm AJ (1990) Heart rate variability. Clin Cardiol 13:570–576

Malik M, Camm AJ (1995) Heart rate variability. Futura Publishing Company, Inc, Armonk

Meregnani J, Clarençon D, Vivier M, Peinnequin A, Mouret C et al (2011) Anti-inflammatory effect of vagus nerve stimulation in a rat model of inflammatory bowel disease. Auton Neurosci 160:82–89

Moga VD, Kurcalte I, Moga M, Vidu F, Rezus C, et al (2014) Chaos theory and non-linear dynamics in hypertensive cardiopathy and heart failure. In: Electrocardiology 2014 – Proceedings of the 41st international congress on electrocardiology, 24, June 4–7, 2014, Bratislava, Slovakia

Montano N, Ruscone TG, Porta A, Lombardi F, Pagani M et al (1994) Power spectrum analysis of heart rate variability to assess the changes in sympathovagal balance during graded orthostatic tilt. Circulation 90:1826–1831
Movahed S, Hermanis E (2008) Fractal analysis of river flow fluctuations. Physica A 387:915–932
Nakamura Y, Yamamoto Y, Muraoka I (1993) Autonomic control of heart rate during physical exercise and fractal dimension of heart rate variability. J Appl Physiol 74:875–881
Natelson BH (1985) Neurocardiology, an interdisciplinary area for the 80s. Arch Neurol 42:178–184
Nepal GB, Paudel BH (2012) Effect of posture on heart rate variability in school children. Nepal Med Coll J 14:298–302
Norris PR, Anderson SM, Jenkins JM, Williams AE, Morris JA Jr (2008) Heart rate multiscale entropy at three hours predicts hospital mortality in 3,154 trauma patients. Shock 30:17–22
Novak V, Novak P, de Champlain J, Le Blanc AR, Martin R et al (1993) Influence of respiration on heart rate and blood pressure fluctuations. J Appl Physiol 74:617–626
Oppenheimer S (2001) Forebrain lateralization and the cardiovascular correlates of epilepsy. Brain 124:2345–2346
Oswiecimka P, Drożdż S, Forczek M, Jadach S, Kwapień J (2014) Detrended cross-correlation analysis consistently extended to multifractality. Phys Rev E 89:–023305
Pachauri N, Mishra DK (2012) Phase synchronization and coherence analysis between ECG & arterial blood pressure. Int J Comput Appl 44:0975–8887
Pagani M, Lombardi F, Guzzetto S, Rimoldi O, Furlan R et al (1986) Power spectral analysis of heart rate and arterial pressure variabilities as a marker of sympathovagal interaction in man and conscious dog. Circ Res 59:178–193
Papaioannou TG, Vlachopoulos C, Loakeimidis N, Alexopoulos N, Stefanadis C (2006) Non-linear dynamics of blood pressure variability after caffeine consumption. Clin Med Res 4:114–118
Parati G, Di Rienzo M, Bertinieri G, Pomidossi G, Casadei R et al (1988) Evaluation of the baroreceptor-heart rate reflex by 24-hour intra-arterial blood pressure monitoring in humans. Hypertension 12:214–222
Pavlov AN, Ziganshin AR, Klimova OA (2005) Multifractal characterization of blood pressure dynamics: stress-induced phenomena. Chaos, Solitons Fractals 24:57–63
Peng CK, Havlin S, Stanley HE, Goldberger AL (1995) Quantification of scaling exponents and crossover phenomena in nonstationary heartbeat time series. Chaos 5:82–87
Pichon A, Roulaud M, Jonville SA, de Bisschop C, Denjean A (2006) Spectral analysis of heart rate variability: interchangeability between autoregressive analysis and fast Fourier transform. J Electrocardiol 39:31–37
Pincus SM (1991) Approximate entropy as a measure of system complexity. Proc Natl Acad Sci U S A 88:2297–2301
Podobnik B, Fu DF, Stanley HE, Ivanov P (2007) Power-law autocorrelated stochastic processes with long-range cross-correlations. Eur Phys J B 56:47–52
Podobnik B, Grosse I, Horvatic D, Ilic S, Ivanov P et al (2009) Quantifying cross-correlations using local and global detrending approaches. Eur Phys J B 71:243–250
Podobnik B, Jiang ZQ, Zhou WX, Stanley HE (2011) Statistical tests for power-law cross-correlated processes. Phys Rev E 84:066118
Pompe B, Blidh P, Hoyer D, Eiselt M (1998) Using mutual information to measure coupling in the cardiorespiratory system. IEEE Eng Med Biol Mag 17:32–39
Porta A, Bassani T, Bari V, Tobaldini E, Takahashi ACM et al (2012) Model-based assessment of baroreflex and cardiopulmonary couplings during graded head-up tilt. Comput Biol Med 42:298–305
Pujitha K, Parvathi G, Sekhar KM (2014) Postural changes in heart rate and blood pressure with ageing. Int J Physiother Res 2:751–756
Pump B, Christensen NJ, Vidbaek R, Warberg J, Hendriksen O et al (1997) Left atrial distension and antiorthostatic decrease in arterial pressure and heart rate in humans. Am J Physiol Heart Circ Physiol 273:H2632–H2638

Pump B, Gabrielsen A, Christensen NJ, Bie P, Bestle M et al (1999) Mechanisms of inhibition of vasopressin release during moderate antiorthostatic posture change in humans. Am J Phys Regul Integr Comp Phys 277:R229–R235

Pump B, Kamo T, Gabrielsen A, Norsk P (2001) Mechanisms of hypotensive effects of a posture change from seated to supine in humans. Acta Physiol Scand 171:405–412

Qian XY, Liu YM, Jiang ZQ, Podobnik B, Zhou WX et al (2015) Detrended partial cross-correlation analysis of two nonstationary time series influenced by common external forces. Phys Rev E 91:062816

Radelli A, Bernardi l, Valle F, Leuzzi S, Salvucci F et al (1994) Cardiovascular autonomic modulation in essential hypertension effect of tilting. Hypertension 24:556–563

Richman JS, Moorman JR (2000) Physiological time-series analysis using approximate entropy and sample entropy. Am J Phys Heart Circ Phys 278:H2039–H2049

Ritzenberg AL, Adam DR, Cohen RJ (1984) Period multupling – evidence for non-linear behaviour of the canine heart. Nature 307:159–161

Saul JP, Berger RD, Albrecht P, Stein SP, Chen MH et al (1991) Transfer function analysis of the circulation: unique insights into cardiovascular regulation. Am J Phys 261:H1231–H1245

Schwab K, Eiselt M, Putsche P, Helbig M, Witte H (2006) Time-variant parametric estimation of transient quadratic phase couplings between heart rate components in healthy neonates. Med Biol Eng Comput 44:1077–1083

Sela R, Hurvich C (2012) The average periodogram estimator for a power law in coherency. J Time Ser Anal 33:340–363

Shoemaker JK, Hogeman CS, Khan M, Kimmerly DS, Sinoway LI (2001) Gender affects sympathetic and hemodynamic response to postural stress. Am J Physiol Heart Circ Physiol 281: H2028–H2035

Sornmo L, Laguna P (2005) Bioelectrical signal processing in cardiac and neurological application. Elsevier Academic Press, Amsterdam

Souza d AC, Cisternas JR, Abreu d LC, Roque AL, Monteiro CBM et al (2014) Fractal correlation property of heart rate variability in response to the postural change maneuver in healthy women. Int Arch Med 7:25–31

Špulák D, Čmejla R, Fabián V (2010) Experiments with blood pressure monitoring using ECG and PPG. In: Proceedings of the conference technical computing Bratislava, pp 1–5

Thayer JF, Yamamoto SS, Brosschot JF (2010) The relationship of autonomic imbalance, heart rate variability and cardiovascular disease risk factors. Int J Cardiol 141:122–131

Thayer JF, Åhs F, Fredrikson M, Sollers JJ III, Wagere TD (2012) A meta-analysis of heart rate variability and neuroimaging studies: implications for heart rate variability as a marker of stress and health. Neurosci Biobehav Rev 36:747–756

Tonhajzerov I, Javorka K, Javorka M, Petrásková M (2002) Cardiovascular autonomic nervous system tests: reference values in young people (15-19 years) and influence of age and gender. Clin Physiol Funct Imaging 22:398–403

Tseng L, Tang S-C, Chang C-Y, Lin Y-C, Abbod MF et al (2013) Non-linear and conventional biosignal analyses applied to tilt table test for evaluating autonomic nervous system and autoregulation. Open Biomed Eng J 7:93–99

Tulppo MP, Mäkikallio TH, Takala TE, Seppänen T, Huikuri HV (1996) Quantitative beat-to-beat analysis of heart rate dynamics during exercise. Am J Physiol Heart Circ Physiol 271:H244–H252

Tulppo MP, Mäkikallio TH, Seppänen T, Laukkanen RT, Huikuri HV (1998) Vagal modulation of heart rate during exercise: effects of age and physical fitness. Am J Physiol Heart Circ Physiol 274:H424–H429

Tulppo MP, Kiviniemi AM, Hautala AJ, Kallio M, Seppänen T et al (2005) Physiological background of the loss of fractal heart rate dynamics. Circulation 112:314–319

Tuzcu V, Nas S, Borklu T, Ugur A (2006) Decrease in the heart rate complexity prior to the onset of a trial fibrillation. Europace 8:398–402

Valenza G, Garcia RG, Citi L, Scilingo EP, Tomaz CA et al (2015) Non-linear digital signal processing in mental health: characterization of major depression using instantaneous entropy measures of heartbeat dynamics. Front Physiol 6:74

Voss A, Schulz S, Schroeder R, Baumert M, Caminal P (2009) Methods derived from non-linear dynamics for analysing heart rate variability. Philos Trans R Soc Am 367:277–296

Wang GJ, Xie C (2012) Cross-correlations between WTI crude oil market and US stock market: a perspective from econophysics. Acta Phys Pol B 43:2021–2036

Wang F, Liao G, Zhou X, Shi W (2013) Multifractal detrended cross-correlation analysis for power markets. Non Linear Dyn 72:353–363

Ward M, ChB MB, Langton JA (2007) Blood pressure measurement. Medscape, 2007

Winter DA, Prince F, Frank JS, Powell C, Zabjek KF (1996) Unified theory regarding A/P and M/L balance in quiet stance. J Neurophysiol 75:2334–2343

Yamamoto Y, Hughson RL (1991) Coarse-graining spectral analysis: new method for studying heart rate variability. J Appl Physiol 71:1143–1150

Zamponi N, Passamonti C, Cesaroni E, Trignani R, Rychlicki F (2011) Effectiveness of vagal nerve stimulation (VNS) in patients with drop-attacks and different epileptic syndromes. Seizure 20:468–474

Zhao X, Shang P, Jin Q (2011) Multifractal detrended cross-correlation analysis of Chinese stock markets based on time delay. Fractals 19:329–338

Zhou WX (2008) Multifractal detrended cross-correlation analysis for two nonstationary signals. Phys Rev E 77:066211

Ziegler D (1994) Diabetic cardiovascular autonomic neuropathy: prognosis, diagnosis and treatment. Diabetes Metab Rev 10:339–383

Appendices

Appendix A

Multifractal Detrended Fluctuation Analysis (MF-DFA)

Multifractal detrended fluctuation analysis was proposed by Kantelhardt et al. (2002) for the study of nonstationary time series which are affected by trends or cannot be normalized. This method, which aims to identify the scaling behavior of the fluctuations of the time series for different qth order moments, is based on the detrended fluctuation analysis (Peng et al. 1994). The method is detailed below:

First let us consider $x(i)$ for $i = 1, \ldots\ldots\ldots, N$, to be a nonstationary time series of length N. The mean of the above series is given by

$$x_{\text{ave}} = \frac{1}{N} \sum_{i=1}^{N} x\,(i) \tag{1}$$

Assuming $x(i)$ to be the increments of a random walk process around the average, the trajectory can be obtained by integration of the signal.

$$Y(i) = \sum_{k=1}^{i} [x(k) - x_{\text{ave}}] \text{for } i = 1 \ldots\ldots N \tag{2}$$

The level of measurement noise present in experimental records and the finite data are also reduced by the integration thereby dividing the integrated time series into N_s nonoverlapping bins, where $N_s =$ int (N/s) where s is the length of the bin. As N is not a multiple of s, a small portion of the series is left at the end. Again, to include that left part, the entire process is repeated in a similar way starting from the opposite end, leaving a small portion at the beginning. Hence, $2N_s$ bins are obtained

D. Ghosh et al., *Multifractals and Chronic Diseases of the Central Nervous System*,
https://doi.org/10.1007/978-981-13-3552-5

altogether, and for each bin least sqaure fit of the series is done followed by determination of the variance.

$$F^2(s, \nu) = \frac{1}{s} \sum_{i=1}^{s} \{Y[(\nu - 1)s + i] - y_\nu(i)\}^2 \tag{3}$$

For each bin ν, $\nu = 1$ .... N_s and

$$F^2(s, \nu) = \frac{1}{s} \sum_{i=1}^{s} \{Y[N - (\nu - N_s)s + i] - y_\nu(i)\}^2 \tag{4}$$

For $\nu = N_s + 1$........, $2N_s$, where $y_\nu(i)$ is the least sqaure fitted value in the bin ν. In our research work, we have performed a least sqaure linear fit (MF-DFA −1). The study can also be extended to higher orders by fitting quadratic, cubic, or higher-order polynomials.

The qth order fluctuation function $F_q(s)$ is obtained after averaging over $2N_s$ bins:

$$F_q(s) = \left\{ 1/2N_s \sum_{\nu=1}^{2N_s} \left[F^2\left(s, \nu\right)^{\frac{q}{2}}\right] \right\}^{1/q} \tag{5}$$

where q is an index which can take all possible values except zero, as the factor $1/q$ becomes infinite with zero value. The procedure can be repeated by varying the value of s. With the increase in the value of s, $F_q(s)$ increases, and for the long-range power correlated series, $F_q(s)$ shows power-law behavior:

$$F_q(s) \propto s^{h(q)}$$

If such a scaling exists, $\ln F_q$ will depend linearly on s with slope $h(q)$. In general, the exponent $h(q)$ depends on q. For a stationary time series, $h(2)$ is identical with the Hurst exponent H. $h(q)$ is said to be the generalized exponent. The value of h (0) cannot be obtained directly, because F_q blows up at $q = 0$. F_q cannot be obtained by normal averaging procedure; instead a logarithmic averaging procedure is applied.

$$F_0(s) \equiv \exp\left\{ 1/4N_s \sum_{\nu=1}^{2N_s} \ln\left[F^2(s, \nu)\right] \right\} \sim s^{h(0)} \tag{6}$$

A monofractal time series is characterized by unique $h(q)$ for all values of q. If small and large fluctuations scale differently, then $h(q)$ will depend on q, or in other words the time series is multifractal. Kantelhardt et al. (2003) have explained that the

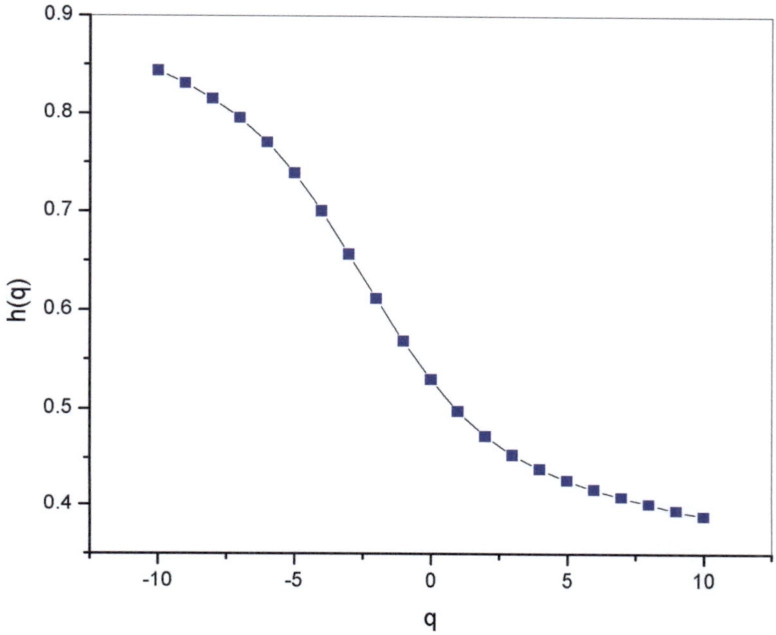

Fig. 1 Plot of $h(q)$ vs. q

values of $h(q)$ for $q < 0$ will be larger than that for $q > 0$. A typical plot of $h(q)$ vs. q is shown in Fig. 1.

The generalized Hurst exponent $h(q)$ of MF-DFA is related to the classical scaling exponent $\tau(q)$ by the relation:

$$\tau(q) = qh(q) - 1 \tag{7}$$

A typical plot of $\tau(q)$ vs. q is shown in Fig. 2.

A monofractal series with long-range correlation is characterized by linearly dependent q- order exponent $\tau(q)$ with a single Hurst exponent H. Multifractal signals have multiple Hurst exponent, and $\tau(q)$ depends non-linearly on q (Ashkenazy et al. 2003a). The singularity spectrum $f(\alpha)$ is related to $\tau(q)$ by Legendre transform (Parisi and Frisch 1985).

$$\propto = \frac{d\tau}{dq} \qquad\qquad f(\propto) = q \propto -\tau(q)$$

where α is the singularity strength or Holder exponent and $f(\alpha)$ specifies the dimension of the subset series that is characterized by α. Using Eq. (7) we can write α and $f(\alpha)$ in terms of $h(q)$

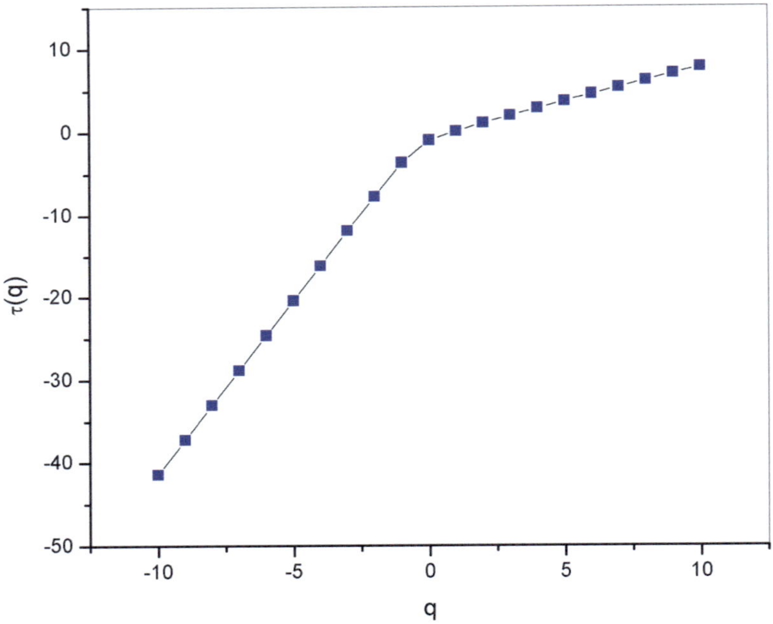

Fig. 2 Plot of $\tau(q)$ vs. q

$$\propto = h(q) + qh'(q) \tag{8}$$

$$f(\propto) = q[\propto - h(q)] + 1 \tag{9}$$

In general, the singularity spectrum quantifies the long-range correlations property of the time series (Ashkenazy et al. 2002). The multifractal spectrum is capable of providing information about the relative importance of various fractal exponents in the time series, e.g., the width of the spectrum denotes range of exponents. A quantitative characterization of the spectra can be done by least square fitting it to quadratic function (Shimizu et al. 2002) around the position of maximum α_0:

$$f(\propto) = A\,(\propto - \propto_0)^2 + B(\propto - \propto_0) + C \tag{10}$$

where C is an additive constant, $C = f(\alpha_0) = 1$, B indicates the asymmetry of the spectrum, and is zero for a symmetric spectrum. The width of the spectrum can be obtained by extrapolating the fitted curve to zero. Width W is defined as $W = \alpha_1 - \alpha_2$ with $f(\alpha_1) = f(\alpha_2) = 0$. It has been proposed by some workers (Ashkenazy et al. 2003b) that the width of the multifractal spectrum is a measure of the degree of multifractality. Singularity strength or Holder exponent α and the dimension of subset series $f(\alpha)$ can be obtained from relations 8 and 9. For a monofractal series, $h(q)$ is independent of q. Hence from relations 8 and 9, it is evident that there will be a unique value of α and $f(\alpha)$, the value of α being the generalized Hurst exponent H

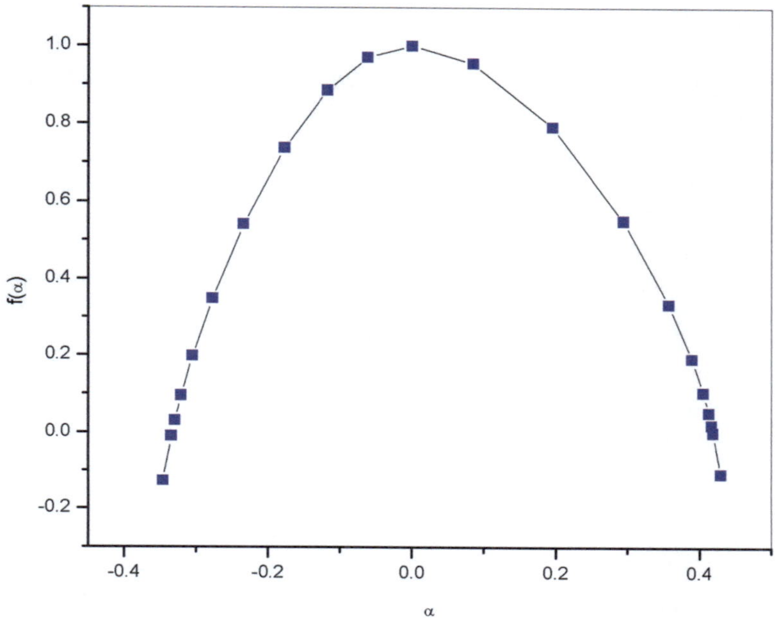

Fig. 3 Plot of $f(\alpha)$ vs. α

and the value of $f(\alpha)$ being 1. Hence the width of the spectrum will be zero for a monofractal series. The more the width, the more multifractal is the spectrum. A sample plot of multifractal spectrum is depicted in Fig. 3.

The auto-correlation exponent γ can be estimated from the relation given below (Kantelhardt et al. 2001; Movahed and Hermanis 2008):

$$\gamma = 2 - 2(h)(q = 2) \tag{11}$$

For uncorrelated or short-range correlated data, $h(2)$ is expected to have a value of 0.5, while a value greater than 0.5 is expected for long-range correlations. Therefore for uncorrelated data, γ has a value of 1, and the lower the value, the more correlated is the data.

Multifractality may be of two types: (i) "due to broad probability density function for the values of time series and (ii) due to different long-range correlation for small and large fluctuation." To ascertain the origin of multifractality, the time series is randomly shuffled and then analyzed. While shuffling the values are arranged randomly so that all correlations are destroyed. The shuffled series will exhibit non-multifractal scaling if multifractality is due to long-range correlation, and if it is due to broad probability density, then, the original $h(q)$ dependence is not changed, $h(q) = h_{\text{shuf}}(q)$. "But if both kinds of multifractality are present in a given series, then the shuffled series will show weaker multifractality than the original one" (Kantelhardt et al. 2002).

Appendix B

Multifractal Detrended Cross-Correlation Analysis (MF-DXA)

In 2008, Zhou (2008), extended the detrended cross-correlation analysis (DXA) method to multifractal detrended cross-correlation analysis (MF-DXA), an advanced version of the DXA method to investigate multifractal behavior between two time series in one or higher dimensions that are recorded simultaneously. The MF-DXA method is a combination of multifractal analysis and detrended cross-correlation analysis and is based on the order detrended covariance (Campillo and Paul 2003; Cottet et al. 2004; Podobnik et al. 2009). Just same as the MF-DFA method, MF-DXA consists of the four steps.

Let us suppose two nonstationary time series $x(i)$ and $y(i)$ for $i = 1\ldots\ldots N$ of length N. The means of the above series' are given by

$$x_{\mathrm{avg}} = \frac{1}{N}\sum_{i=1}^{N} x(i) \quad \& \quad y_{\mathrm{avg}} = \frac{1}{N}\sum_{i=1}^{N} y(i) \tag{12}$$

The profiles of the underlying data series $x(i)$ and $y(i)$ are computed as

$$X(i) \equiv \sum_{k=1}^{i}\left[x(k) - x_{\mathrm{avg}}\right] \text{for } i = 1\ldots\ldots N.$$

$$Y(i) \equiv \sum_{k=1}^{i}\left[y(k) - y_{\mathrm{avg}}\right] \text{for } i = 1\ldots\ldots N \tag{13}$$

The integration also reduces the level of measurement noise present in experimental records and finite data. Each of the integrated time series is divided to N_s nonoverlapping bins where $N_s = \mathrm{int}(N/s)$ where s is the length of the bin. Now since N is not a multiple of s, a short part of the series is left at the end. So in order to include this part of the series, the entire process is repeated starting from the opposite end thus leaving a short part at the beginning thus obtaining $2N_s$ bins. For each bin, least sqaure linear fit is performed, and the fluctuation function is given by

$$F(s,\nu) = \frac{1}{s}\sum_{i=1}^{s}\{Y[(\nu-1)s+i] - y_\nu(i)\} \times \{X[(\nu-1)s+i] - x_\nu(i)\}$$

for each bin $\nu, \nu = 1, \ldots\ldots N_s$ and

$$F(s,\nu) = \frac{1}{s}\sum_{i=1}^{s}\{Y[N-(\nu-N_s)s+i]-y_\nu(i)\}\times\{X[N-(\nu-N_s)s+i]-x_\nu(i)\}$$

for $\nu = N_s+1, \ldots \ldots 0.2N_s$, where $x_\nu(i)$ and $y_\nu(i)$ are the least sqaure fitted values in the bin ν.

The qth order detrended covariance $F_q(s)$ is obtained after averaging over $2N_s$ bins.

$$F_q(s) = \left\{\frac{1}{2N_s}\sum_{\nu=1}^{2N_s}[F(s,\nu)]^{q/2}\right\}^{1/q} \tag{14}$$

where q is an index which can take all possible values except zero because in that case the factor $1/q$ blows up. The procedure can be repeated by varying the value of s. $F_q(s)$ increases with increase in value of s. If the series is long-range power correlated, then $F_q(s)$ will show power-law behavior:

$$F_q(s) \propto s^{\lambda(q)}$$

If such a scaling exists, ln F_q will depend linearly on ln s, with $\lambda(q)$ as the slope. Scaling exponent $\lambda(q)$ represents the degree of the cross-correlation between the two time series. In general the exponent $\lambda(q)$ depends on q. We cannot obtain the value of $\lambda(0)$ directly because F_q blows up at $q = 0$. F_q cannot be obtained by the normal averaging procedure; instead a logarithmic averaging procedure is applied:

$$F_0(s) \equiv \exp\left\{\frac{1}{4N_s}\sum_{\nu=1}^{2N_s}\ln[F(s,\nu)]\right\} \sim s^{\lambda(0)} \tag{15}$$

For $q = 2$ the method reduces to standard DXA.

$F(s,\nu)$ may obtain negative values in general. To eliminate the problem in evaluation of fluctuation functions which may be complex valued for different values of q, we have taken the modulus of $F(s,\nu)$ to eliminate the negative values. However, Oswiecimka et al. (2014) proposed an alternative more rigorous method multifractal cross-correlation analysis (MFCCA) to take care of the negative values in cross covariances. The authors suggest that the proposed method is a more natural generalization of DCCA compared to MF-DXA. It prohibits losing information that is stored in the negative cross-covariance. The method is yet to be tested in various systems.

If scaling exponent $\lambda(q)$ is independent of q, the cross-correlations between two time series are monofractal; on the other hand if $\lambda(q)$ depends on q, the cross-correlations between two time series are multifractal. Furthermore, for positive q, $\lambda(q)$ describes the scaling behavior of the segments with large fluctuations, and for negative q, $\lambda(q)$ describes the scaling behavior of the segments with small fluctuations. Scaling exponent $\lambda(q)$ represents the degree of the cross-correlation between the two time series $x(i)$ and $y(i)$. The value $\lambda(q) = 0.5$ denotes the absence of cross-correlation. $\lambda(q) > 0.5$ indicates persistent long-range cross-correlations where a large value in one variable is likely to be followed by a large value in another variable, while the value $\lambda(q) < 0.5$ indicates anti-persistent cross-correlations where a large value in one variable is likely to be followed by a small value in another variable and vice versa (Movahed and Hermanis 2008; Shadkhoo and Jafari 2009).

Zhou (2008) found that for two time series constructed by binomial measure from p model, there exists the following relationship between scaling exponent and Hurst exponent:

$$\lambda(q = 2) \approx \left[h_x(q = 2) + h_y(q = 2)\right]/2 \tag{16}$$

Podobnik and Stanley have studied the above relation for monofractal autoregressive fractional integral moving average (ARFIMA) signals and EEG time series (Podobnik and Stanley 2008; Shadkhoo and Jafari 2009). Zhou has shown that the above relation holds for any q for multifractal random walks (MRW) and binomial measures generated from the p model (Mars and Lopes da Silva 1983; Hajian and Movahed 2010). However, there are also examples in which the above relation does not exist for all values of q, such as daily price changes for DJIA and NASDAQ indices (Podobnik and Stanley 2008; Shadkhoo and Jafari 2009), but for $q = 2$ it is still correct (Podobnik and Stanley 2008; Shadkhoo and Jafari 2009). The other example is the case of two time series generated by using two uncoupled ARFIMA processes, each of both is auto-correlated, but there is no power-law cross-correlation with a specific exponent (Podobnik and Stanley 2008; Shadkhoo and Jafari 2009).

According to auto-correlation function given by

$$C\,(\tau) = \langle[x(i+\tau) - \langle x\rangle][x(i) - \langle x\rangle]\rangle\tilde{}\tau^{-\gamma} \tag{17}$$

Hajian and Movahed (2010) introduced the cross-correlation function as

$$C_x(\tau) = \langle[x(i+\tau) - \langle x\rangle][y(i) - \langle y\rangle]\rangle\tilde{}\tau^{-\gamma_x} \tag{18}$$

where γ and γ_x are the auto-correlation and cross-correlation exponents, respectively. Due to the non-stationarities and trends superimposed on the collected data, direct calculation of these exponents is usually not recommended; rather the reliable method to calculate auto-correlation exponent is the DFA method, namely, $\gamma = 2 - 2h$ $(q = 2)$ (Kantelhardt et al. 2001; Movahed and Hermanis 2008). Podobnik and Stanley (2008) have demonstrated the relation between cross-correlation exponent, γ_x, and scaling exponent $\lambda(q)$ derived by Eq. (15) according to $\gamma_x = 2 - 2\lambda(q = 2)$. For uncorrelated data, γ_x has a value of 1, and the lower the value of γ and γ_x, the more correlated is the data.

In general, $\lambda(q)$ depends on q, indicating the presence of multifractality. In other words, we want to point out how two series are cross-correlated in various time scales. To clarify this correlation, we generalize the singularity spectrum, $f(\alpha)$, concept to two cross-correlated series. This generalized concept gives useful information about the distribution of the degree of cross-correlation in different time scales. The way to characterize multifractality of cross-correlation between two series is to relate via a $\lambda(q)$ Legendre transform, as in the case of one series (Peitgen et al. 1992; Wang et al. 2012).

$$\alpha = \lambda(q) + q\lambda'(q) \tag{19}$$

$$f(\alpha) = q[\alpha - \lambda(q)] + 1 \tag{20}$$

Here, α is the singularity strength or Hölder exponent, while $f(\alpha)$ denotes the dimension of the subset of the series that is characterized by α. Unique Hölder exponent denotes monofractality, while in the multifractal case, the different parts of the structure are characterized by different values of α, leading to the existence of the spectrum $f(\alpha)$. The width of the spectrum can be obtained by extrapolating the fitted curve to zero. Width W is defined as

$$W = \alpha_1 - \alpha_2 \tag{21}$$

with $f(\alpha_1) = f(\alpha_2) = 0$. The growth of the width of $f(\alpha)$ shows the increase in the degree of multifractality of two coupled signals.

References

Ashkenazy Y, Hausdroff JM, Ivanov PC, Stanley HE (2002) Stochastic model of human gait dynamics. Physica A 316:662–670

Ashkenazy Y, Havlin S, Ivanov PC, Peng CK, Frohlinde VS et al (2003a) Magnitude and sign scaling in power-law correlated time-series. Physica A 323:19–41

Ashkenazy Y, Baker DR, Gildor H, Havlin S (2003b) Nonlinearity and multifractality of climate change in the past 420,000 years. Geophys Res Lett 30:2146–2149

Campillo M, Paul A (2003) Long-range correlations in the diffuse seismic coda. Science 299:547–549

Cottet A, Belzig W, Bruder C (2004) Positive cross correlations in a three-terminal quantum dot with ferromagnetic contacts. Phys Rev Lett 92:206801

Hajian S, Movahed MS (2010) Multifractal detrended cross-correlation analysis of sunspot numbers and river flow fluctuations. Physica A 389:4942–4957

Kantelhardt JW, Koscielny BE, Rego HHA, Havlin S, Bunde A (2001) Detecting long-range correlations with detrended fluctuation analysis. Physica A 295:441–454

Kantelhardt JW, Zschiegner SA, Koscielny BE, Havlin S, Bunde A et al (2002) Multifractal detrended fluctuation analysis of nonstationary time series. Physica A 316:87–114

Kantelhardt JW, Rybski D, Zschiegner SA, Braun P, Bunde EK et al (2003) Multifractality of river runoff and precipitation: comparison of fluctuation analysis and wavelet methods. Physica A 330:240–245
Mars NJ, Lopes da Silva FH (1983) Electroencephalogr Clin Neurophysiol 56:194–209
Movahed S, Hermanis E (2008) Fractal analysis of river flow fluctuations (with Erratum). Physica A 387:915–932
Oswiecimka P, Drożdż S, Forczek M, Jadach S, Kwapień J (2014) Detrended cross-correlation analysis consistently extended to multifractality. Phys Rev E 89:023305
Parisi G, Frisch U (1985) Turbulence and predictability in geophysical fluid dynamics and climate dynamics. Elsevier Science Ltd., Amsterdam/New York
Peitgen HO, Jürgens H, Saupe D (1992) Chaos and Fractals. Springer, New York 1992, Appendix B
Peng CK, Buldyrev SV, Havlin S, Simons M, Stanley HE et al (1994) Mosaic organization of DNA nucleotides. Phys Rev E 46:1685–1689
Podobnik B, Stanley HE (2008) Detrended cross-correlation analysis: a new method for analyzing two nonstationary time series. Phys Rev Lett 100:084102
Podobnik B et al (2009) Proc Natl Acad Sci U S A 106:22079
Shadkhoo S, Jafari GR (2009) Multifractal detrended cross-correlation analysis of temporal and spatial seismic data. Eur Phys J B 72:679–683
Shimizu Y, Thurner S, Ehrenberger K (2002) Multifractal spectra as a measure of complexity in human posture. Fractals 10:103–116
Wang J, Shang PJ, Ge WJ (2012) Multifractal cross-correlation analysis based on statistical moments. Fractals 20:271–279
Zhou WX (2008) Multifractal detrended cross-correlation analysis for two nonstationary time series. Phys Rev E 77:066211

MIX
Papier aus verantwortungsvollen Quellen
Paper from responsible sources
FSC® C105338

If you have any concerns about our products,
you can contact us on
ProductSafety@springernature.com

In case Publisher is established outside the EU,
the EU authorized representative is:
Springer Nature Customer Service Center GmbH
Europaplatz 3, 69115 Heidelberg, Germany

Printed by Libri Plureos GmbH
in Hamburg, Germany